S. S. Rakhimkhodjaev
G. N. Sobirova

Processes of raw material preparation for weaving

S. S. Rakhimkhodjaev
G. N. Sobirova

Processes of raw material preparation for weaving

ScienciaScripts

Imprint

Any brand names and product names mentioned in this book are subject to trademark, brand or patent protection and are trademarks or registered trademarks of their respective holders. The use of brand names, product names, common names, trade names, product descriptions etc. even without a particular marking in this work is in no way to be construed to mean that such names may be regarded as unrestricted in respect of trademark and brand protection legislation and could thus be used by anyone.

Cover image: www.ingimage.com

This book is a translation from the original published under ISBN 978-620-7-48786-8.

Publisher:
Sciencia Scripts
is a trademark of
Dodo Books Indian Ocean Ltd. and OmniScriptum S.R.L publishing group

120 High Road, East Finchley, London, N2 9ED, United Kingdom
Str. Armeneasca 28/1, office 1, Chisinau MD-2012, Republic of Moldova, Europe
Printed at: see last page
ISBN: 978-620-8-16507-9

Contents

OUTLINE.

The work shows modern technologies and equipment of preparatory operations of warp and weft. The following processes are considered: winding and winding of yarns on the wadding during rewinding, weaving and sanding.

The special place is given to the problem of thread tension in the process of raw material preparation for weaving and fabric production, as well as ways of their stabilisation. For researchers, technologists, designers, masters and bachelors engaged in the processes of raw material preparation for weaving.

INTRODUCTION

The prerequisite for the origin of weaving was the availability of raw materials. At the weaving stage, these were strips of animal skin, grasses, reeds, etc. Woven clothes and shoes, mats, baskets and nets were the first weaving products.

Weaving predates spinning because weaving existed before man discovered the spinning ability of animal and plant fibres such as cotton, nettles, flax, hemp, wool, down, etc.

The manufacture of fabric is a very labour-intensive operation, as it requires the preparation of a group of warp yarns and a continuous weft yarn.

Initially, hand weaving used circular filling of warp between two bars, the distance between which was made equal to the length of the woven fabric and to give the threads of the warp smoothness made sanding with a special composition.

The thread was wet before winding on the stick (shuttle) to make it soft and smooth.

The development of machine (mechanical) weaving required the separation of the warp and weft preparation process and its transfer to machine technology.

At the beginning of the 19th century the first sanding and winding machines appeared, and by the middle of the 19th century the first spinning machines appeared.

Modern machines (equipment) of the preparation department are computerised and allow the production of semi-finished products (bobbins, rolls, navoi, etc.) of high quality.

Taking into account the fact that Uzbekistan has a powerful raw material base (cotton, silk, wool, chemical fibre), the government pays great attention to the production of finished national products (fabrics, clothes, etc.) and saturation of both domestic and foreign markets.

1.REWINDING OF YARNS AND THREADS

1.1.Yarn winding

The input bundle in yarn rewinding is usually the cob, whose shape and structure depends on the ratio of the ring (ring spinning machine) diameter to the winding diameter. The structure of the cob must ensure a high yarn winding speed.

In the process of cob actuation from the toe to the socket, the thread tension increases 15 - 20 times. This is due to: yarn balance in the cylinder; forces acting on the ballooning yarn; shape and size of the cylinder; cob winding density; cob taper; arrangement of yarn turns in the winding taper; cob size; chuck size.

Yarn equilibrium in the cylinder means an equilibrium winding in which there is no slippage of the yarn coils from the cob surface during the yarn winding process. The winding of the yarn on the cob shall be carried out on a geodetic line. The main indicator determining the shape and structure of the cob is the geodesic deflection angle Θ, i.e. the angle between the normal N to the winding surface at point M and the main normal F at the same point lying in the contact plane Q (Fig. 1).

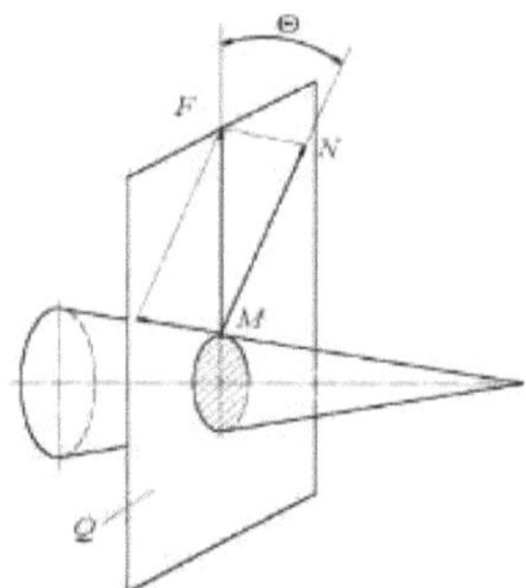

Fig.1.To calculate the angle of geodetic deviation.

The critical (highest) value of the geodesic deflection angle 0_{κ} is the main factor determining the area of winding equilibrium, within which winding of the yarn on the bundle can be carried out.

The value of 0k depends on the following factors:
-   thread tension during cob formation;
-   the nature and size of the pack;

-   the nature of the fibre;
-   linear yarn density;
-   the speed of the thread winding;
-   temperature and humidity conditions in the shop;
-   curing time and moisture content of the yarn.

The critical angle of geodesic deflection $\theta_к$, at which the filament is at the threshold of breaking from the winding surface, is obtained from the expression

$$\text{tg } \theta_к \leq f_{max}$$

where: f_{max} is the maximum coefficient of friction of fibrous materials.

With maximum friction angle (φ_{max}) dependence

$$\theta_к \leq \varphi_{max}$$

To assess the conditions of equilibrium winding, it is necessary to determine the actual value of the geodesic deviation angle $0_п$, and then compare it with the value of the critical angle $0_к$. If the calculated angle $0_п$ is close to the angle $0_к$, then the winding on the pack is not sufficiently equilibrium.

The calculated value of angle $0p$ is determined by the formula:

$$\text{tg } \theta_п = \text{tg } \alpha(1+\sin^2 \gamma) / \cos \gamma$$

where: α - angle of packer cone, degree; γ - angle of winding lift, degree.

The forces affecting the tension of the thread in the cylinder include:

-   the pre-tension of the yarn obtained during cob formation;
-   the friction force of the thread against the cartridge;
-   the force caused by the acceleration of the mass at the winding point when the thread is brought out of rest.

The equation of the thread tension at any point of the cylinder has the form:

$$T_е = T_о (\exp f\alpha) + m\, v_с^2 + (m\omega^2/ 2)\, (R^2 - r^2)$$

where: $T_о (\exp fa)$ - tension of the thread winding off the cob; m - mass of the ballooned thread; $V_с$ speed of separation of the thread from the cob; R - radius at the point of winding off the cob; ω - angular velocity of the ballooned thread; r - radius of the balloon.

The tension of the thread at the top of the cylinder at $r=0$ is:

$$T = T_о (\exp f\alpha) + mv_с^2 + m\omega^2 R^2 / 2$$

The first and second expressions are the tension of the yarn at the point of separation from the bundle, and the third expression is the dynamic component of tension. The amount of thread tension in the cylinder determines the shape and dimensions of the cylinder.

The cylinder shape (number of waves) increases with increasing linear winding speed and filling distance from the cob to the yarn guide. It has been observed that at the beginning of the winding up of the yarn from the full cob there is a multiwave winding up of the yarn.

As the cob is triggered, especially in the cob nest, the balloon becomes single-wave, which leads to a sharp increase in thread tension and, as a consequence, to thread breakage. To prevent this phenomenon during rewinding, balloon breakers are used, which ensure multi-wave rewinding regardless of the place where the thread is rewound from the cob.

The length of the yarn in a cob of the same volume is determined by the winding density. The higher the winding density on the packer, the less slippage (better winding of the yarn) when unwinding the yarn. The winding density is determined by the yarn tension, which, if exceeded, leads to a deterioration of the physical and mechanical properties of the yarn.

On the other hand, increasing the winding density increases the yarn rewinding speed. The cob taper, defined by the ratio of the cob cone height to the winding diameter, has a significant influence on the rewinding speed, the higher it is, the better is the yarn winding process.

The yarn turns are arranged on the winding cone as a ply (slow movement of the ring slat on the spinning machine) and a layer (fast reverse movement of the slat). A slow slat movement in the layer forms a parallel winding, while a fast slat movement with a large pitch forms a cross winding. This arrangement of coils causes a sudden change in balloon shape and thread tension. To equalise the thread tension, it is advisable to use a plyless winding. In the case of layerless winding the turns of different directions are crossed evenly, the pitch of the layer turns increases, the pitch of the layer turns decreases, which leads to the formation of cross winding.

The winding of yarns of different linear densities from different cob sizes starts with a multi-wave and ends with a single-wave cylinder. The larger the diameter of the cartridge and the outer diameter of the winding, the slower the decrease in the number of waves in the cylinder. With increasing winding height and diameter, winding cone height, yarn linear density, the conditions for winding the yarn off the cob deteriorate.

1.2 Winding the yarn

In order to produce a coiled bundle, at least two motions must be imparted to the yarn - transfer and translational. This is possible by means of a winding and spreading mechanism in the form of a winding drum. The type, shape, structure and capacity of the winding bundle depend on the following factors: the pitch of the helical line and the angle of inclination of the coil; the tension of the yarns; the specific winding density; the angle of shift of the coils; the position of the bobbin axis in relation to the axis of the winding drum.

Consider each of the factors. The thread is applied to the surface of the bobbin along a helical line. The nature of the arrangement of thread coils on the forging depends on the pitch of the helical line and the angle of inclination of the coil. The angle of inclination of the coil determines the crossing angle of the coils of the winding, which ranges from 30 to 55 degrees. The crossing angle of 55 degrees is mainly used to obtain bobbins with minimum specific winding density for dyeing. For subsequent weaving processes, bobbins with a crossing angle of 30 degrees are used, in which the maximum specific winding density on the packet is created. The value of the angle of crossing of turns depends on the number of turns on the winding drum, which can be 1; 1.5; 2 and 2.5 turns. With 1 revolution, the drum cycle is two revolutions, and with 2.5 revolutions it is five revolutions. For weaving production, in order to improve the conditions of winding, reels with a constant winding cone angle are used, in which the angle of inclination of the coil increases from the large diameter to the small diameter of the reel, and for knitting and twisting production, and vice versa.

Obtaining the correct bundle structure depends on the tension of the yarns during the rewinding process, which is created by the tensioning devices. On modern winders, special compensators are installed to improve the control of thread tension, reacting to any changes in thread tension and adjusting the tensioner both per cycle of the reel movement and for the entire bobbin run.

The yarn tension during rewinding, the crossing angle, the bobbin pressure on the reel, the linear yarn density influence the specific winding density on the bobbin.

Let's consider one element of crossed thread segments in two layers of the bobbin (Fig.2). The volume occupied by the fibre material without taking into account the buckling of the thread is equal to:

$$V = 2 \cdot d \cdot l_1 \cdot l_2; \qquad l_1 = l \cdot \sin\alpha \; ; \qquad l_2 = l\cos\alpha$$
$$V = 2 \cdot d \cdot l^2 \sin\alpha \cdot \cos\alpha = d \cdot l^2 \sin 2\alpha;$$
$$2\sin\alpha \cdot \cos\alpha = \sin 2\alpha \; ;$$

Subject to buckling

$$V = d \cdot l^2 \sin 2\alpha \cdot K$$

where: K - buckling coefficient, usually equal to 0.6 - 0.9

Weight of a piece of thread

$$m = \frac{2 \cdot l}{N} = \frac{2 \cdot l \cdot T}{1000}$$

where: N - yarn number; T - linear yarn density.

Specific gravity

$$\gamma = \frac{m}{V} = \frac{2 \cdot l \cdot T}{1000 \cdot d \cdot l^2 \sin 2\alpha \cdot K} = \frac{2 \cdot T}{1000 d \cdot \sin 2\alpha \cdot K}$$

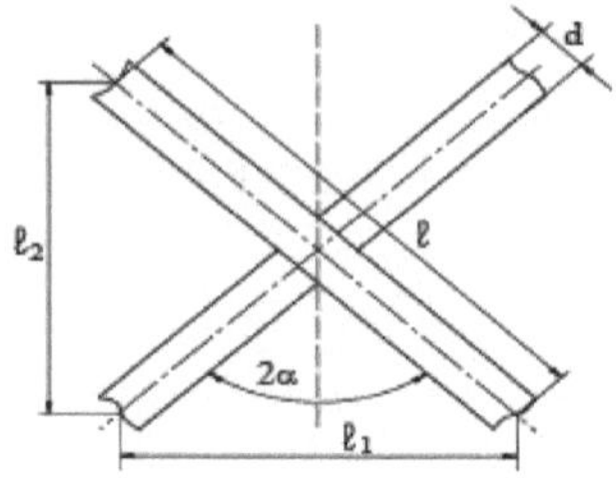

Figure 2. Schematic of the crossed sections of thread in the two layers of the bobbin.

The specific winding density is inversely proportional to the sine of the crossing angle of the turns. The lowest winding density is at $2a = 90^0$ and the highest winding density at the lowest crossing angle, i.e. parallel winding. Due to the elasticity of the yarn, the yarn tension of each winding turns acts radially on the layers (turns) below. Let us determine the value of the specific pressure of one strand of yarn on the bundle. On the surface of the bundle let's select an elementary section of a coil with length (Fig. 3).

From the equilibrium condition of the coil segment

$$\frac{dQ}{2} = \frac{t \cdot \sin d\varphi}{2}$$

$$dQ = \frac{2t \cdot \sin d\varphi}{2}$$

If we assume that the angle is small, it can be taken with a small error

$$\frac{\sin d\varphi}{2} = \frac{d\varphi}{2}.$$

In this case the pressure of the elementary section of the coil

$$dQ = \frac{2t\sin d\varphi}{2} = \frac{2t \cdot d\varphi}{2} = t \cdot d\varphi$$

Specific pressure value of the coil

$$q = \frac{dQ}{dl} = \frac{t \cdot d\varphi}{(R \cdot d\varphi)} = \frac{t}{R}$$

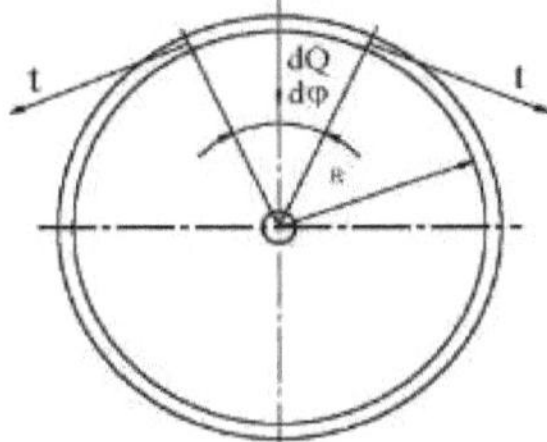

Fig. 3. Calculation of the specific pressure of the coil on the packer.

$$dl = R \cdot d\varphi$$

It follows from the equation that the sealing pressure of a coil per unit length is proportional to the tension and inversely proportional to the winding radius.

Due to the pressure of the outer winding layers, the inner layers are compacted and somewhat compressed towards the centre of the bundle. As a result, the tension of the coils in the inner layers is reduced. The tension is maintained in the outer layers and in the inner layers directly lying on the base of the bundle. During the rewinding process, the mass of the bobbin gradually increases, hence the pressure of the bobbin on the reel increases, resulting in a change in the winding density. In order to maintain a constant pressure of the bobbin on the reel during the rewinding process, it is advisable to provide the machines with systems for controlling and stabilising the pressure of the bobbin on the reel.

In practice, the winding density is determined with a device called a densimeter. Table 1 shows approximate values for the specific winding density depending on the yarn type and the winding on the bundle.

Table 1

Specific winding density values depending on yarn type and winding on the bundle.

№	Type of yarn	Specific winding density, g/cm	
		Type of winding on the pack	
		Parallel winding	Cross winding
1	Cotton	0,45-0,5	0,35-0,45
2	Linen	0,5-0,7	0,43-0,5
3	Artificial silk	0,5-0,6	0,58-0,72

| 4 | Natural silk | 0,5-0,6 | 0,58-0,72 |

The value of specific winding density at the ends of the bundle is 1.5-2 times higher than in the middle part of the reel. At the same time, the winding at the larger end is more dense than at the smaller end, this is due to the position of the bobbin axis relative to the drum axis.

As can be seen from Fig. 4, when the axes of the bobbin and the drum are parallel, the friction forces acting on the ends of the bobbin will be directed perpendicular to its axis.

Let us decompose the tension force K in two mutually perpendicular directions into the force K_1, which ensures proper winding density, and the force K2, which is absorbed by the edge of the drum groove.

At point A, the friction force acts in the opposite direction, as the bobbin lags behind the drum at this point. This results in negative pressure and braking torque on the bobbin.

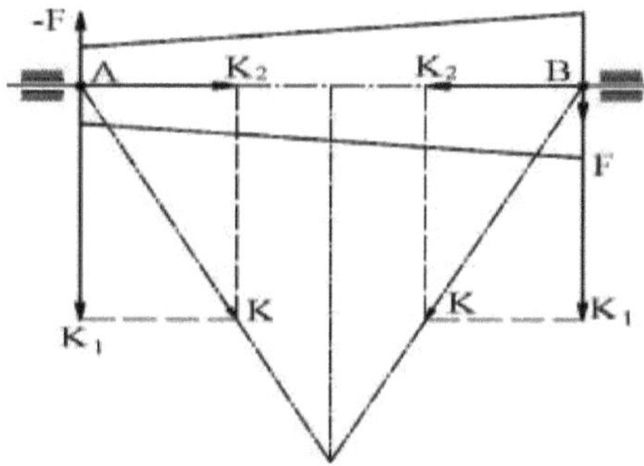

Figure 4. Diagram of friction forces acting on the ends of the spool.

The total force determining the winding density is equal:

$$K_1 + (-F) = K1 - F$$

At point B, the friction force F is reversed, so the total force determining the winding density is already equal:

$$K_1 + (+F) = K1 + F$$

Partial elimination of this disadvantage is possible by shifting the axis of the bobbin relative to the axis of the drum by 2-3 degrees anti-clockwise (Fig. 5), as well as setting a tight clamping of the small end of the bobbin to the drum and a gap of 1.5-2 mm between the large end of the bobbin and the drum

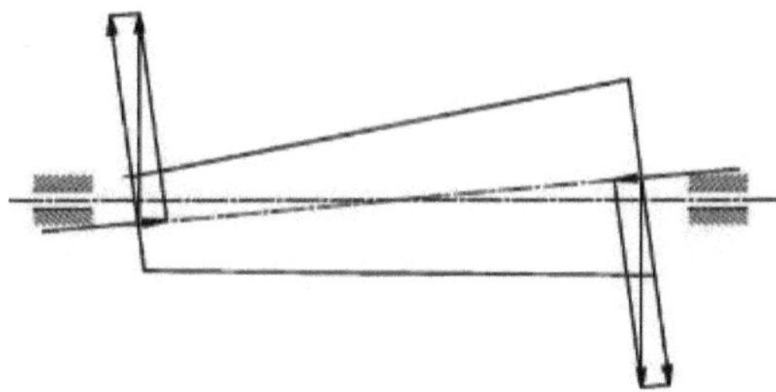

Figure 5. Schematic of bobbin axis displacement relative to the drum axis.

A homogeneous structure of the bobbin can be obtained when the places of change of direction of the threads are evenly distributed over the whole surface of the bobbin. However, at certain winding diameters on the bobbin, the threads overlap each other in the form of a ribbon (bundle), i.e. there is an overlap of the tops of the coils in one place. This indicates that there is no shift angle of the coil tops.

Tape (harness) winding is a major vice.

Due to the uneven surface, the bobbin is shaken violently at the moment of tourniquet formation, which causes axial displacement of the bobbin. This causes the threads to fly off the end of the bobbin.

When winding up the yarn from the bobbin, the bundles cause massive yarn breakage.

A bundle winding during the thread winding process is only formed when the diameter of the bobbin is equal to or a multiple of the diameter of the reel.

To prevent bundle winding, the bobbins are given, in some cases, a variable motion and in other cases, a rocking or reciprocating motion in the axial direction.

It is advisable to install the bobbin holders on the right when winding Z twist yarns and on the left when winding S twist yarns. This prevents uneven winding and yarns from flying off the large bobbin end.

1.3. Yarn inspection and cleaning

Packs coming from the spinning mill have a number of defects on the yarn:

1. Nodules - vegetable impurities in the fibres, small thickenings. The reason for this is poor raw material quality and faulty fibre processing. Difficulty in removing this defect reduces the yarn quality.

2. Adjacent - have a corkscrew appearance, the thickenings are due to work technique, it is not possible to remove them without a knot.

3. Scale - formed by a build-up of fibres on the yarn. Loosely bound yarn impurities are removed without thread breakage, while rigidly bound yarn impurities are removed with thread breakage.

4. Bumps are caused by improper operation of the drafting devices on roving and spinning machines. It is not possible to eliminate the defect without a knot.

5. Cracks - caused by insufficient yarn clamping in the drafting unit.

6. Unspun threads are thickenings of the thread itself that have a slight twist. They are removed by breaking the thread.

7. Thick threads - double threads, large thickenings, poor roving piecing. Removal of this defect is very difficult.

Visual evaluation of yarn defects is very difficult, therefore the "Uster-Klassimat" system is used, in which yarn defects are classified according to diameter and length and divided into sixteen groups (Fig. 6).

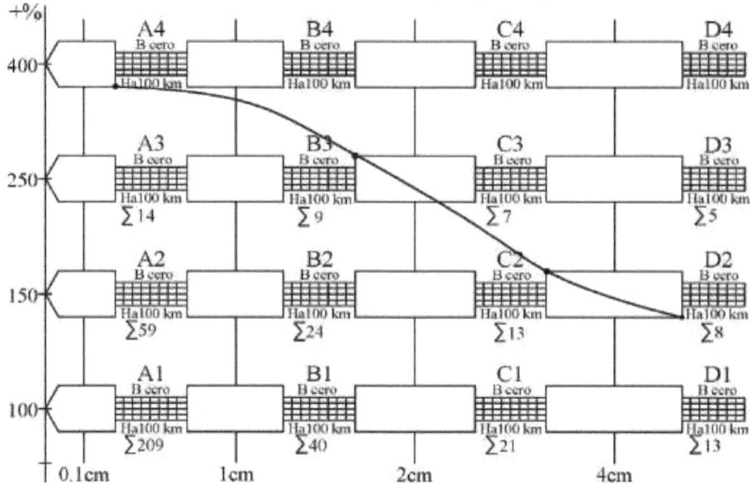

Fig.6. Scheme of the system "Uster-Klassimat."

The four groups A, B, C and D correspond to malformations lengths of 0.1, 1, 2 and 4 cm, respectively, and the sensitivity in per cent shows the percentage increase in the cross-sectional area of the malformations from + 80 to +400 %, corresponding to an increase in diameter from 60 to 150 %. As a result, we have sixteen groups from A1 (the shortest in length and smallest in diameter malformations) to D4 (the longest and largest in diameter malformations).

In the tables of each classification group, the number of defects is recorded: in the upper line for a certain sample length, in the lower line for 105 m of yarn.

Group A and B includes yarns with thin and thick places. This is the area of normal deviations due to random fibre distributions.

Group C includes yarn defects such as lumps, cracks, plaques and piecing defects.

To group D - non-spreading, large thickenings, double threads.

Fig. 6 shows an example of the analysis of a polyester/cotton yarn (67/33 %)

according to the Uster-Classimat system. The suitability of a yarn for a particular purpose is determined by the international standards for the presence of defects in the yarn type, or by the fabric quality requirements. If similar yarn tests are available that have not been cleaned, the accumulated frequencies in each vice length group for the specified yarn are plotted and compared.

Irrespective of the setting of any yarn cleaning limits, the D4 and D3 vices are definitely removed.

The yarn cleaning curve is then plotted at the specified sensitivity at the vicious length using the Uster correlator (Fig. 6).

As can be seen from Fig. 6, the defects of groups A4, B4, C4, D4, D3, C3 and part of groups A3 and B3 should be eliminated, while the remaining defects are considered acceptable for this fabric.

It can be calculated that in 10^5 metres of yarn there will be 23 knots in the eliminated groups (half in the boundary group A3 and B3).

Electronic yarn cleaning and the use of the Uster-Classimat system make it possible to produce fabrics with an acceptable level of yarn defects with high yarn quality and minimum processing costs.

There are two methods of electronic yarn cleaning, capacitive and opto-electronic, where there is no contact between the yarn and the working parts of the device.

Capacitive cleaners from Qualitex and Uster measure the mass of a unit length, deviation from a given mass will cause a capacitive change, which in turn will change the reference voltage which will drive the knife that cuts the thread.

A formula is recommended for assessing yarn quality:

$$F = (Nc / Ny)\ 100\ \%$$

where: Nc - number of eliminated defects during cleaning;

Ny - number of knots in the cleaned yarn.

It must be borne in mind that during yarn processing, a knot tied in place of a cut-out blemish (short, thick section) can cause worse consequences than a leftover blemish. In weaving, about half of all yarn breaks are caused by knots and the costs for removing them during the finishing process are also increased.

The Uster opto-electronic cleaners make it possible to take a photo of a blemish at any time to determine and compare the yarn quality according to the Uster-Klasimat system and to remove blemishes with the selected device setting.

Electronic cleaners ensure a uniform yarn cleaning degree despite variations in moisture content, line density, chemical additives, blend ratios and differences between blend components.

1.4. Joining the ends of the threads

The knotting of yarn ends is a mass operation performed by the spinning and weaving workers. Therefore, carelessly made knots (weak knots with large tendrils) break off, which leads to weaver's workload and increased weaving machine downtime.

The distribution of main yarn breakage on the weaving machine by cause is given in the work of Uster and TNO in Table 2.

Table 2

№	Causes of discontinuity	TRO data, %	Uster's data. %	
			Factory No. 1	Factory No. 2
1	Thin places	6		
2	Pooh	1147	22	18
3	Unknowns	30		
4	reasons	2153	65	488
5	Nodes	32	78	2
	Thickenings		13	34

The analysis of Table 2 shows that the increase in yarn breaks is due to loosely tied knots. Excessive yarn cleaning increases the number of knots, which worsens the weaving process.

It is better to optimise yarn cleaning than to use a device that can detect any defect and replace it with a knot. According to a study by Garten and Moyler, 8-10 % of all knots come undone during the weaving process.

Let's assume that the production of fabric Mitkal with a density of 25 n/cm on the base, 22 n/cm on the weft, with a linear density of yarn on the base and weft 20 tex, on the machine STB-330 in three webs width of 100 cm, speed 220 min^{-1} KPV = 0.9 is organised.

Let's determine the output of the machine:

A = (220 x 60 / 220 x 10) x 3 x 0.9 = 16.2 m/hour

Number of metres of warp yarns processed on the machine per 1 hour

L = 16.2 x 25 x 100 = 40500 m

Number of nodes per 1 hour of machine operation r = 40500 x 50 / 10^5 = 20.25 nodes

Where: 50 / 10^5 is the maximum number of knots per 100 km of a given yarn.

Taking into account that on average 9% of the knots will be weak, we will get (20.25 x 9) / 100 = 1.8 stops due to knot loosening for 1 hour of machine operation. On 1 m of fabric breakage due to knot loosening will be 1,8/16,2=0,11 breaks. The norm of breakage for this assortment is 0.15 breaks per 1 m of fabric. The obtained value of yarn breakage in weaving only because of the knot is very high, so it is necessary to reduce the number of cleaning knots in the yarn, in our example, somewhere in two times (25 knots per 100 km of yarn).

The physical and mechanical properties of the knot are greatly influenced by the type of knot. The strongest and most resistant to shock and vibration are self-tightening knots (fisherman's knots) and double knots (figure-eight knots). From the point of view of knot passability in weaving and workability into the fabric - one-breadth (weaving) and self-tightening knots. To obtain quality knots, a special device called a knotter is used, which can be manually or automatically knotted.

Manual and automatic knotters are categorised by number.

If a knotter designed for thin yarns is used with thick yarns, a knot with short ends will result and will quickly come undone.

If a thick yarn knotter is used for thin yarns, a knot with long ends will result, which will cause additional friction, clinging of the knot section to the loom parts and to the neighbouring warp threads during weaving.

Piecing is also used to join the yarn ends. The piecing places increase the yarn diameter by 20 % and the tensile strength of this yarn section is at least 80 % of the tensile strength of normal yarn.

The piecing is carried out with adhesives such as carboxyl methyl cellulose (CMC), polyacrylamide (PAA), polyvinyl alcohol (PVS) to strengthen the yarn ends.

Piecing can be manual (done manually by workers) or mechanical on special devices (splicers), which are fitted on modern winding machines. The splicers are installed individually on each winding head.

With a pulse of compressed air, the splicer twists the torn yarn ends by loosening the ends beforehand and makes a knotless connection. A standard splicer makes this connection.

For twisted yarns of vegetable origin, injection splicers are used, where liquid is added to the air supplied to join the yarn ends, which increases the strength of the fibre connection.

For yarns made of animal fibres and their blends with synthetic fibres, thermal splicers are used, where the air supplied to join the yarn ends is heated, which ensures better fibre fixation.

The quality of the splice is checked with an electronic cleaner. Schlafhorst and Murata winding machines are equipped with splice systems.

1.5.Rewinding equipment.

At present, winding machines have replaced winding machines in many enterprises. However, the use of coiling machines is useful for rewinding:

- large bundles, twisted, worsted complex and textured yarns;
- pressure dyeing bobbins;

- The wefts working in very difficult conditions of weft insertion into the shed on needleless machines;
- from cylindrical bobbins in rotor spinning to conical bobbins;

with simultaneous trotting of the yarn for twisting.

Precision winding with specific density from 0.25 to 0.7 g/cm^3 , with rewinding speed from 700 to 1000 m/min, forced rewinding of the input bundle, emulsifying device (for charge removal) is characteristic for winding machines and automatic machines of "Scherer", "Georg SACM", "Zilbos", "Schlafhorst", "Murata".

It is known that rotor spinning yarns have the best unevenness in short lengths and the worst unevenness in long lengths, which is why they are rewound for fabrics of a certain quality. The Uster-Klasimat tests show that the yarn has a higher unevenness on long lengths in groups D4 and D3. As the number of defects in these groups is low and their elimination by electronic cleaners is not difficult, the efficiency of rewinding from rotor spinning reels to conical reels is obvious.

The modern level of winding machines is characterised by the following features - improved quality of winding and unwinding of bundles; increased productivity of labour and equipment; optimal use of resources; easy operation and maintenance; high safety in operation.

Improved winding and unwinding of the bundles is achieved through yarn tension control by means of special tensioning systems; optimum yarn running conditions; continuous yarn and rewinding monitoring; improved smooth yarn acceleration and uniform bundle winding. The increase in productivity is due to rewinding speeds of up to 2000 m/min, optimised unwinding of the yarn from the bundles and economical yarn cleaning.

Optimum resource utilisation is achieved by minimising yarn waste and reducing energy consumption through the electronic rewind control system.

High operational safety, ease of operation and maintenance are ensured by the modular design of the units, electronic controls, monitoring and setting of the main rewinding parameters. Every winding head consists of an unwinding unit, a centre unit and a winding unit, which facilitates operation, adjustment, cleaning and maintenance. The elements of the unwinding unit ensure reliable feeding and positioning of the bundle and optimum conditions for the yarn to come off the bundle. The elements of the central unit ensure yarn and bundle quality through constant yarn quality control, yarn tension control and yarn splice. The winding unit elements ensure smooth start-up and acceleration of the reel, stepless rewinding speed, uniform winding on the bundle, damping and weight compensation of the bobbin holder. In addition, in order to prevent big end tears and uneven winding, the bobbin holders are installed on the right side for Z twist yarns and on the left side for S twist yarns.

In the process of rewinding, the yarn passes through:

- yarn take-off accelerator, which ensures an increase in yarn tension uniformity over the entire cob trip period;
- loop brake to prevent twisting when starting the winding head; - pre-cleaner;
- tensioning device or yarn tension control system;
- electronic cleaner for yarn quality control and for checking yarn splice points;
- yarn cutting, clamping and catching device during yarn splicing. - automatic yarn splicer (splicer).

The winder also has a power unit, an information system, a fluff and dust removal system, a bobbin remover and a cob and bobbin inlet and outlet conveyor.

Further improvement of winding machines and automatic machines is carried out in the following directions:

- increasing the degree of automation of rewinding;
- development of new electronic systems for control and regulation of rewinding process parameters;
- modernisation and unification of the main units, devices and mechanisms.

The increase in the degree of automation of the rewinding process is centred on the integration of automatic winding machines with spinning machines (winding and spinning unit). The automatic transport and sorting system according to yarn type enables the rewinding of several types of processed yarns at the same time. The new control systems with computers allow to monitor and evaluate the quality of yarn joining, length and density of the yarn on the bobbin, tension changes, optimise the yarn rewinding process and provide operational information.

Unification of parts and assemblies means their interchangeability. Development of perfect systems of tension control, as the cob is triggered and winding on the bobbin, systems of finding and feeding the yarn of the required length into the splicer, systems of yarn splicing, yarn quality control, systems preventing sliver (bundle) winding, systems of dumping behind the big end of the bobbin and yarn winding on the reel, systems of smooth stop and acceleration of the reel, systems providing the set density and working out of the running length of the yarn on the bobbin allows to improve the rewinding process and to increase the quality of the output bundles. The increase of the equipment utilisation efficiency is possible with the transition to larger sizes of input and output bundles, with the reduction of yarn breakage, with the installation of a knotting machine on each winding head.

(piecing) station, when equipping the cob tray feeder and bobbin remover, when converting idle time into productive time. Fig. 7 shows the efficiency of the machines depending on the knotting stations per number of winding heads.

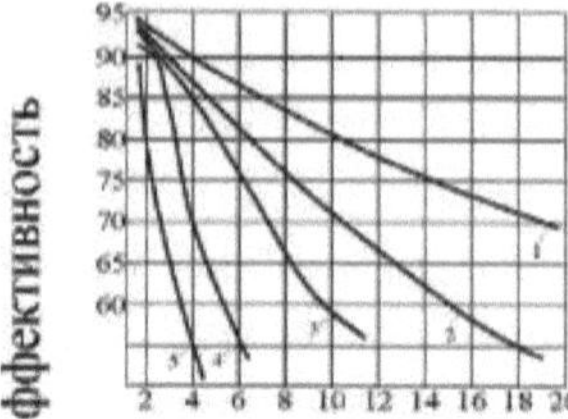

Number of connections per

10,000m of yarn

Figure 7. Machine efficiency as a function of knotting stations per number of winding heads.

1   -one knotting (piecing) station per one winding head
2   -one knotting station for 5 winding heads
3   -one knotting station for 10 winding heads
4   -one knotting station for 16 winding heads
5   -one knotting station for 24 winding heads

Fig. 7 shows that the efficiency of the winding machines increases when the yarn breakage is reduced and the number of winding heads operated by the knotting station is reduced.

The required number of workers per unit of output, if we assume 100% when equipping the coiler with a shop of cob stock and without an autoloader, and when equipping a store-type automatic machine with an autoloader will be 85%, then for a coiler with continuous feeding of cobs (pallet type) without an autoloader - 35%, and when equipping a pallet-type automatic machine with an autoloader - 21%. There is an average 5-fold reduction in the need for workers when using a pallet-type automatic machine with an auto-lifter.

The conversion of downtime into productive time means a reduction in the deceleration time of the winding reel acceleration, a reduction in the time required to find and join the yarn ends, an increase in the speed of bobbin changing by the bobbin remover, simplified machine maintenance (modular construction, easy access to all machine components, quick centralised input of the main parameters via the information system, keeping the surrounding area clean).

1.6 Yarn properties

During rewinding, the yarn is subjected to tensile and friction forces. This causes some changes in the physical and mechanical properties of the yarn.

After rewinding, the yarn linear density decreases, elongation decreases, strength remains unchanged. The loss of elongation and decrease in yarn linear density is due to the drawing in of the yarn during rewinding. Also the loss in yarn mass due to cleaning during rewinding leads to a decrease in yarn line density. When rewinding yarns from fixed forgings, a change in twist occurs, i.e. the twist of the yarn increases by one twist when one twist comes off the spinning cob. It is determined by the following relationship:

$$K = K_1 + \frac{1}{\pi \cdot d}$$

where: K_1 is the yarn twist before rewinding per one metre of yarn;
d - average cob diameter, m.
As can be seen, the small cob diameter causes a significant increase in twist.
The degree of change in the physical and mechanical properties of the yarn during rewinding depends on the rewinding speed and the yarn tension. The filling tension is set at 5-10% of the yarn strength.
As the rewinding speed increases, the filling tension of the yarns decreases. If the friction forces arising during the rewinding process are weak, the yarn becomes frayed (abrasion) and may even break. It is therefore necessary to set up the correct tram line for the yarn and its interaction with the machine's working elements.
Special yarn treatments such as scorching, emulsifying, oiling, paraffinising and waxing are also carried out to improve the technological properties of some yarns.
Scorching reduces yarn hairiness and reduces yarn breakage in weaving. It is used for yarns made of various fibres, including synthetic fibres.
In the preparation of chemical fibres for rewinding, emulsifying, oiling and paraffinising are used to improve yarn smoothness, resulting in reduced friction, lower yarn tension and better processing. Yarns made of cotton, viscose fibres and cotton/viscose blends are subjected to waxing, which reduces the breakage of waxed weft yarns by an average of 40 % during rewinding and processing on the weaving machine. The composition of emulsions, greasers, paraffin and waxes should include substances that can be easily removed from the waxed fabrics. The technological process of rewinding should be organised in such a way that the physical and mechanical properties of the yarn change minimally. The criterion for assessing the technological properties of the yarn is the yarn breakage. The causes of yarn breaks during rewinding are numerous and varied. The factors causing yarn breaks can be permanent, periodic or random.
The constant character of the breakage is the spinning vices of the yarn. The number of breaks is determined during the yarn cleaning process by

predetermined defect sizes according to the Uster-Klasimat system, and the number of breaks during yarn rework on the cob, depending on the cob weight. More cob weight, less breaks during cob change for a fixed bobbin weight.

The periodic nature of the breakage is a change in the yarn winding and rewinding parameters. This is mainly influenced by the yarn tension during rewinding. For example: when winding the yarn in the cob socket, the tension changes sharply due to additional friction on the chuck body; the ballooned yarn has a high angular velocity, which leads to flying of the coils from the cob cone. Reduction of breakage is possible by stabilising the yarn tension through measures to improve the winding of the yarn from the cob and by modernising the devices and systems involved in yarn rewinding. For example: installation of a freely rotating cob holder or giving the ballooning yarn an additional rotation in the direction of its rotation in the winding cylinder.

Accidental breakage is caused by yarn dressing geometry irregularities, machine malfunctions, wear and burrs on the yarn line, winding faults on the cob, etc.

As you can see, the difficulties in determining the causes of thread breaks are different - some are the result of carelessness and accident and are easily eliminated, others require some effort, to eliminate the third need to find ways, and the fourth with the existing technique and technology is impossible to eliminate at all.

1.7 Weft yarn rewinding

The difference between warp and weft rewinding is that the size ratio of the feed and exit wadding changes. In warp rewinding machines, the feeding of the input packs must be automated, while in weft rewinding machines, the output packs must be automated.

At a yarn winding speed of 25 tex, a 30 g weft bobbin is wound in 15 minutes and even less for thicker yarns, which results in frequent bobbin changes.

The weft bobbin winding on the weaving machine also causes a frequent change of used bobbins in the same time. This is why weft yarns with a higher linear density are rewound onto packages without a rigid base, i.e. tubular cobs, and in some cases the weft winding head "Unifil" is used on weaving machines where a small number of bobbins ensure a continuous weft supply to the machine.

The filling thread is rewound onto a bobbin of a certain size and construction. The structure of the bobbin winding must ensure the stability of the bobbin when braking the shuttle. When braking the bobbin weighing 30 g. acts on the winding force of 50 N and more, which is quite enough to fall off the coils in the wrong structure of winding. The bobbin should have the maximum diameter with a minimum location from the wall of the shuttle. It should also have an upper and lower reserve winding zone. The upper zone of reserve winding makes it easier to find the end of the thread, and the lower zone of reserve winding prevents the work of the fabric with the defect "knocked down pattern" when working with weft stylus. In order to avoid damage to the winding when changing bobbins provide for orientated - neat stacking of full bobbins. Let us consider some technological characteristics of the weft rewinding process. Equilibrium and winding density can be achieved by the appropriate package shape (cone angle, differential winding), yarn tension and yarn friction coefficient (depends on the yarn type). As the thread tension increases, the geodesic deflection angle decreases sharply up to some limit, after which there is little change. It is recommended to rewind cotton, woollen, linen yarn at a thread tension of at least 20 cN., and artificial silk at least 15 cN. For better yarn convergence it is necessary that the winding taper angle of the bobbin winding is considerably smaller than the critical deflection angle, whereby for smooth yarns (artificial silk) the winding taper angle should be smaller than for rough yarns. The optimum bobbin winding taper for cotton yarns is $26\text{-}28^0$, woollen yarns $28\text{-}32^0$, artificial silk $16\text{-}20^0$. With differential winding there is a scattering of the winding tops, which increases the bobbin winding density by 30 % and reduces the probability of winding off and winding down of the bobbins. The bobbin winding density ranges from 0.3 to 0.35 g/cm^3 (for thick machine yarns) to 0.75 g/cm^3 (for synthetic and combed yarns). On the bobbin, the winding density is achieved by changing the yarn tension and on tubular cobs by the yarn tension and the pressure of the cone former on the winding. The value of the weft thread tension during rewinding is set at 5-15% depending on the tensile strength of the thread. Excessive tension worsens physical and mechanical properties, increases breakage during rewinding.

Constant winding diameter. A constant winding diameter can be achieved by constantly moving the winding zone, for this purpose yarns with a uniform diameter are used. The constant displacement of the winding zone results in a better bobbin structure, but the bobbin shape is often uneven in diameter (lumpy). The yarn is not subjected to the additional influences of the stylus roller used in the variable winding centre movement. In the variable winding zone movement, the bobbin has a strictly cylindrical shape, but the winding structure is worse, especially when the bobbin slot is formed.

The thread length in the winding layer. When producing fabrics on machines with different widths, the yarn length in the layer must be adjusted from 5 to 15 metres. The use of yarns with a higher yarn density on wide machines requires special elements to increase the yarn length in the ply, e.g. the height of the yarn stacker cam and the adjustment of the yarn stacker arm. Upper and lower reserve winding. The upper end of the thread should be short and tuckable, while the lower end should be the shortest length possible and brought to the winding surface. These characteristics affect the productivity of the workers. Direction of rotation. The rotation of the spindle can impart or eliminate twist on the yarn. The number of twists by which the twist of the yarn is reduced or increased, which has been communicated to it in spinning at a certain rate, is not desirable. During winding, if the yarn has a twist S, the number of twists increases and if it has a twist Z, the number of twists decreases. The variation of the number of twists is up to 5 % of the total specified yarn twist.

Rewinding speed. The rewinding speed V is the sum of the forward speed V_1 and the variable speed V_2 :

$$V_1 = \pi \cdot d \cdot n_1 \qquad V_2 = 2h \cdot n_2$$

where: d - average diameter of yarn winding on the package, m; n_1 - number of bobbin revolutions, $\min^{-1}$; h - spread of the yarn guide, m; n_2 - number of revolutions of the eccentric, which gives motion to the yarn guide, $\min^{-1}$.

$$V = \sqrt{V_1^2 + V_2^2} = \sqrt{(\pi \cdot d \cdot n_1)^2 + (2 \cdot h \cdot n_2)^2}$$

Depending on the machine design, fibre type, type and structure of the bundle, the optimum weft rewinding speed is selected.

The Unifil weaving head provides complete automation of the weft feeding of weaving machines. The head consists of a winding spindle with spindle stacker, a reserve winding mechanism, a reserve mechanism for six full bobbins and the feeding of empty bobbins to the spindle. The used bobbins are cleaned from the residual bobbins and transported in an orientated manner by means of magnets to the spindle.

Advantages: Reduced production space, reduced number of chargers, no bobbin cleaning costs; reduced use of bobbins to 18 per machine.

2.YARN WEAVING
2.1 Preparation of the base

The parallelised yarn group is wound up from input bundles containing yarns with few defects. Re-winding is usually used to remove defects. If the number of yarns per warp is small, weaving is carried out directly from the input bundles. A large number of yarns requires an intermediate accumulation in the form of a roll or drum. In all cases, bobbins are used in which the input bundles are placed. Fig. 8 shows the warp preparation variant depending on the number of yarn defects, the size of the input bundles and the physical and mechanical properties of the yarn.

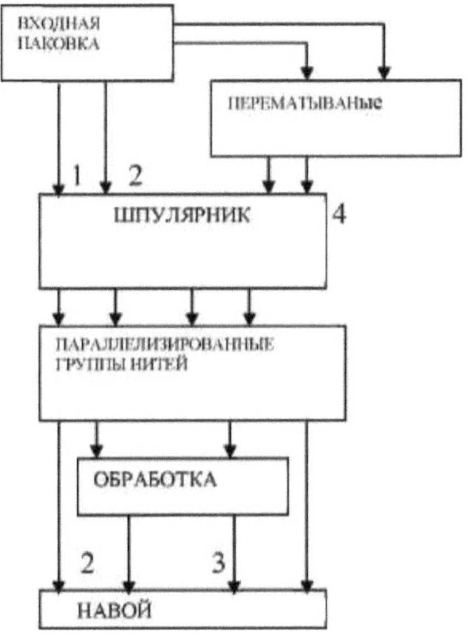

Figure 8. Variants of base preparation.

Variant 1 comprises two processes: yarn parallelisation and warp formation, which is possible in the case of yarns free of blemishes, large sized bundles and high physical and mechanical yarn properties (twisted yarns).

Option 2 is possible for rotor spinning yarns (low physical and mechanical properties) that require treatment to protect them from breaking during weaving.

Variant 3 requires rewinding and processing. Rewinding is necessary to clean the yarn of defects or due to the small size of the entry packs. Processing protects the yarn (unspun cotton yarns) from deterioration during weaving.

In variant 4, rewinding is used and processing is excluded due to the high strength of the yarn (natural silk).

During processing (sanding, emulsifying), the yarn can be twisted from warp to warp, from the spinning roll or rolls to the warp, from the drum to the warp, from the creel to the warp. Table 3 gives recommendations for the choice of the type of knitting.

Table 3

Recommendations on the choice of the type of weaving.

№	Factor to be considered	Direct winding of the warp	Partitioning	Ribbon weaving
1.	Number of shafts (navoi)	One	One or more depending on the number of threads and warp length	One
2.	Total number of threads	Limited by creel capacity	Any, but the number of threads in each roll is the same according to the bobbin capacity	Any
3.	Combination colours	Any	Limited simple strips defined by the number of shafts Limited simple strips defined by the number of shafts	Anything but repeated according to the ribbons
4.	Manual labour costs	Minor	Minor	large
5.	Scope of application	Speciality, carpet backings	Cotton and staple fabrics	Universal application

The combination of colours in the warp is not difficult with ribbon weaving and direct warp to warp. Difficulties arise in batch weaving. The main challenge is to find the easiest way to distribute the coloured threads on each roll of the batch and to determine the bobbin rate in the bobbin case.

The thread count of the coloured threads in a sewing pattern is equal to the sum of the number of threads of all the colours in the pattern.

The total colour pattern is the smallest number of coloured threads, after which the order of their alternation in the pattern is repeated. Partial, I_{qv} colour pattern is the pattern of coloured threads per spinning roller.

$$R^I_{цв} = \frac{R_{цв}}{K_в}$$

where: $K_в$ is the number of spinning rolls.

In the calculation of weaving with coloured bases, there can be four distributions of partial colour patterns on the backing rollers.

The first case is that the coloured threads are evenly distributed on each roller.

This is the simplest case, in which the individual rapports on all the spinning rollers are equal to each other $(R^I_{mв\,1} = R^I_{mв2} = \ldots = R^I_{mвi})$ and the rate is the same.

Example: Prepare a multicolour warp with the number of main threads n_o = 1824 of which edge threads $n_{кp}$ = 24. Colour range $R_{ÜB}$ = 30 threads, red = 15, blue = 9, white = 6 threads, number of needle rollers $K_в$ = 3.

1- Determine the number of filament yarns on each roll

$$n_в = \frac{n_o - n_{кp}}{K_в} = \frac{1824 - 24}{3} = 600$$

2 Determine the number of edge threads on each needle roller

$$n^1_{кp} = \frac{n_{кp}}{K_в} = \frac{24}{3} = 8$$

Consequently, there will be 600 background threads and 8 edge threads on each roller. The distribution of coloured threads on the rollers should be shown in Table 4.

Table 4.

Distribution of coloured threads on the rollers.

Colour range in the fabric	Number of threads each colour	Number of threads on the spinning roller		
		First	second	third
red	15	5	5	5
blue	9	3	3	3
white	6	2	2	2
Bottom line:	Qcv = 30	K'tsv = 10	K'tsv = 10	K'tsv = 10
Number of repeats of the pattern 60	1800	600	600	600
edges	24	8	8	8
Total:	1824	608	608	608

The second case is that the coloured threads are distributed unevenly on each spinning roller, but without missing any of the colours of the total pattern and with a mandatory equality of the individual patterns. In this case, several variants of the distribution of coloured threads on each roller are possible. In this case, it is necessary to strive to reduce the number of different bets during the preparation of the entire batch.

Example: prepare a multicolour base with a number of 1944 main threads, of which 24 threads for the edges, alternating colours in a total rapport of 60 red, 8 blue and 28 white, colour rapport $AND_{u1!}$ = 60+8+28 = 96 threads. The number of spinning rollers $к_v = 3, 648$ threads in each roller. Let us show the distribution of threads in the form of Table 5.

Table 5

Distribution of coloured threads on the rollers

Colour range in the fabric	Number of threads of each colour	Number of threads on the reamer roller		
		first	The second	third
red	60	20	20	20
blue	8	4	4	4
white	10	10	10	10
Bottom line:	96	32	32	32
Repeat 20 times	1920	640	640	640
edges	24	8	8	8
Total:	1944	648	648	648

Table 5 shows the rational method of warp preparation, as the rate on the bobbin is changed at least twice. In other layouts, if the private rapports are equal, the rate is changed three times.

The third case - coloured threads on each roll are distributed unevenly with skips of some colours with obligatory equality of private colour rapports.

Various options for the distribution of coloured threads are possible, but it is necessary to find a rational option with the least number of rate changes.

Example: Prepare a multi-coloured base with 1540 main yarns and 40 edge yarns. The alternation of threads in the total rapport is 10 red, 5 blue, 6 white and 9 yellow. Rapport of colour $AND_{ full} = 10+5+6+9=30$ threads. The number of mending rollers $K_в = 5$ with 308 threads on each roller. The distribution of coloured threads is shown in Table 6.

Table 6.

Distribution of coloured threads on the rollers

Colour range in the fabric	Number of threads of each colour	Number of threads on the spinning roller				
		first	second	third	fourth	fifth
red	10	2	2	2	2	2
blue	5	1	1	1	1	1
white	6	-	-	-	3	3
yellow	9	3	3	3	-	-
Bottom line:	30	6	6	6	6	6

26

Repeat 50 times	1500	300	300	300	300	300
edges	40	8	8	8	8	8
Total:	1540	308	308	308	308	308

Two bobbin stakes are required to weave a multi-coloured warp.

Fourth case. The coloured threads are distributed on the roller by colour. Each colour is whisked onto a separate roller. The need of the rolls is as many as the number of colours of warp threads.

Example; prepare a multicolour base with a number of 2544 main threads, of which 24 edge threads. The alternation of threads in the total rapport is 14 red and 7 blue, colour rapport Kcv = 14 + 7 = 21 threads. Renewal can be done on 6 rolls of 420 threads each for the background and 4 threads each for the edges. Since there are twice as many red threads as blue threads, it is possible to distribute 4 rolls with red threads and two rolls with blue threads. The distribution of the coloured threads is shown in Table 7.

Table 7.

Distribution of coloured threads on the rollers.

Colour range in the fabric	Number of threads of each colour	Number of threads on the spinning roller					
		1 - om	2 - om	3 - м	4 - om	5 - om	6 - om
red	14	14	14	14	14	-	-
blue	7	-	-	-	-	7	7
Bottom line:	21	420	420	420	420	420	420
Repeat the pattern		30	30	30	30	60	60
edges	24	4	4	4	4	4	4
Total:	2544	424	424	424	424	424	424

For ribbon weaving, the number of threads in the ribbon (stakes) must be equal to the whole number of rapports of the colour.

Number of ribbons in the base

$$K_{_\pi} = \frac{n_o}{n_{_\pi}} = \frac{n_o}{K \cdot R_{_{\text{цв}}}}$$

where: po - number of threads in the warp; pl - number of threads in the ribbon; K_π - number of repetitions

the colour pattern in the ribbon.

The number of threads in the ribbon (rate) is equal to the whole number of colour rapports, in order to avoid disturbance of the colour pattern.

For patterning medium-length warp yarns, Karl Mayer offers the new Multi-

Matic patterning machine, which is very efficient and covers warp yarns up to 1500 metres long. It works with a drum patterning mechanism that allows up to 128 threads to be fed onto the back frame. The core of the machine is formed by 128 laying pins, which are radially arranged along the circumference of the drum and carry out, by means of a linear drive, very fast and precise thread laying with a lift of up to 400 mm for the corresponding large cone chuck. Based on the classic yarn feed, the Multi-Matic frame processes yarns made of different fibre types, ranging from silk and natural fibre yarns to complex yarns made of almost any yarn with no limitation in yarn tension. The two sensor-controlled and motor-adjustable yarn brakes are located on the rewinding frame: Multitens for the narrow tension range of 3-220 cN and RotoTens for the wide tension range of 30-550 cN. The latter can be used for high line density yarns with yarn direction via the yarn guide roller, which operates with a braking force of 80 cN.

2.2. creels

The creel must have different properties depending on the area of use, the type and size of the input lugs, the material to be processed, the size of the production area and the selected technology.

Modern bobbins are simple and at the same time stable in construction. The following principles are used in the construction of the creel - the individual elements are assembled in unitised sections and the combination of several such sections forms a complete creel frame. If necessary, the creel can be shortened or lengthened.

Depending on the application, bobbins are classified as follows;
- by appearance (shape) of the creels (rectangular and wedge-shaped);
- by the type of thread weaving process (interrupted or continuous thread weaving);
- according to the type of winding of the yarn from the bundle (parallel or perpendicular to the axis of the bundle);
- by bobbin capacity (capacity);
- by the pitch (spacing) between the packs;
- by the type of bundles to be rewound.

Rectangular bobbins can be standard, trolley bobbins, bobbins with turning sections and shop bobbins.

The standard bobbin case is equipped with a double-sided fixed bobbin holder frame 1 (Fig. 9).

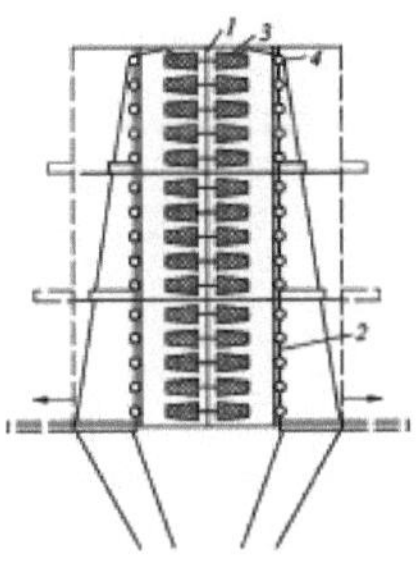

Figure 9. Standard creel.

The movable frames 2 of the thread tensioners are installed with the possibility of adjusting the centredness and the distance between the bobbins 3 and the tensioners 4. The system provides easy access to the bobbins and tensioners.

The bobbin carrier (Fig. 10) is characterised by a double-sided bobbin carrier frame 1 consisting of several bogie sections 2, which can be individually extended or pushed into the bobbin carrier. The bobbins are loaded onto the trolley sections outside the creel (in the winding section) and can hold from 50 to 150 bobbins. The trolley sections can be moved easily by means of running rollers. While one trolley section is winding up the thread, the other trolley section outside the creel is being threaded. This creel design significantly reduces the downtime of the rewinding machine and increases its efficiency.

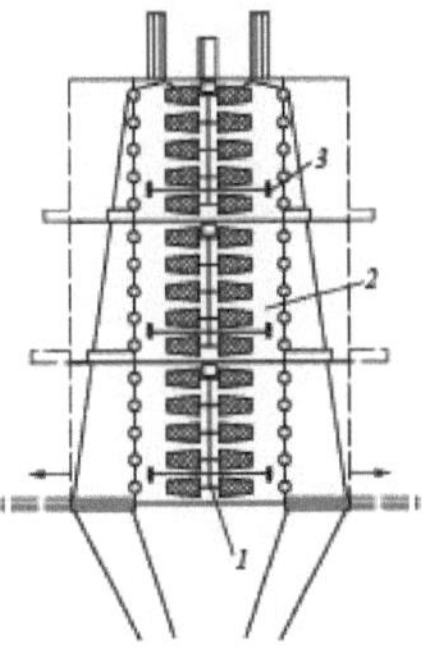

Fig. 10: Trolley creel.

It is advisable to use a trolley bobbin in case of space constraints.

The bobbin with rotary section (Fig. 11) is mainly designed for processing large input bundles of 5 to 25 kg in weight. The swivelling section is made double-sided. On one side the threads are unwound from bobbins 1, and on the other side new bobbins 2 are threaded.

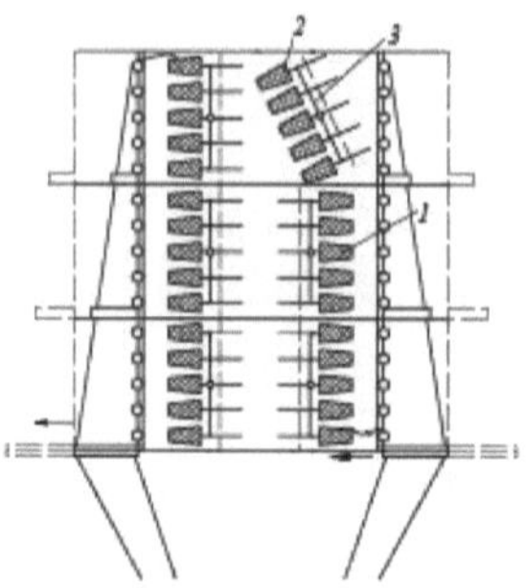

Fig. 11. Bobbin with turning section.

Each section 3 is rotated by 180 degrees with respect to the axis during refuelling and its precise fixation with respect to the movable frames of the thread tensioner. It is advisable to fill large and heavy bundles outside the bobbin, it creates convenience in maintenance and efficient use of auxiliary means when changing bundles. If the bundles have less weight, it is possible to turn the bobbin 180 degrees around the axis of each pair of bobbin holders when re-filling the spinning machine.

The magazine bobbin (Fig. 12) is economically feasible when sewing longer warp lengths. For each thread tensioner 1, a working bobbin 2 and a reserve bobbin 3 are filled, with the beginning of the thread from the reserve bobbin connected to the end of the working bobbin. Replacement of used bobbins are carried out during the operation of the spinning machine. It is expedient to use shop bobbins for weaving of complex yarns from large cobs, which have minimum thread breakage and occupy a small production area, as well as such yarns for which fillings (residue) of yarns on the remaining bundles are highly undesirable.

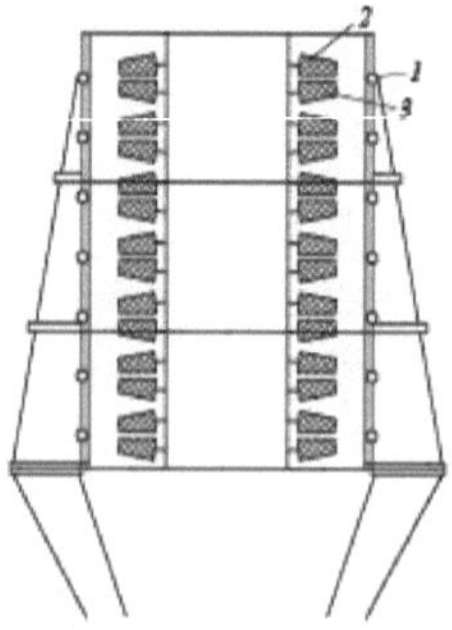

Fig.12. Shop creel.

In the wedge-shaped bobbin, in contrast to the described parallel section bobbins, both wings of the sections are placed at an angle to each other, forming a V-shaped structure (Fig. 13). The inner space 1 between the V-shaped wings of the creel is used for storing and filling the bobbins when changing batches.

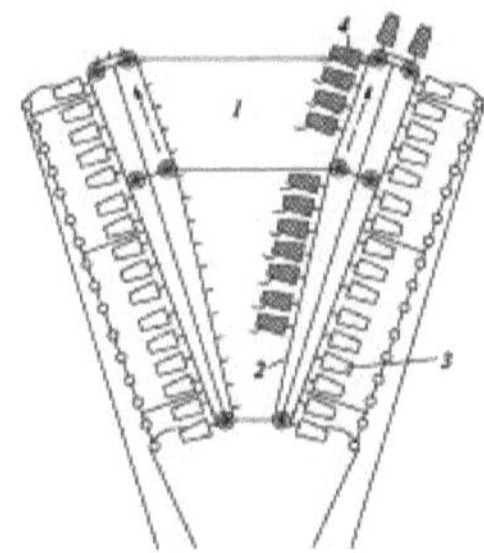

Fig. 13. Wedge-shaped creel.

The bobbin-holding frames are designed as an endless chain 2. While one bobbin rate 3 is being threaded from the outside of the bobbin case, the next bobbin rate 4 is threaded onto the empty spindles placed on the inside of the bobbin case. When changing the rate, the empty bobbins are transferred to the inner side of the bobbin case and the full bobbins to the outer side of the bobbin case.

All types of bobbin weavers are equipped to a greater or lesser extent with bobbin side displacement, fans, additional stop and start buttons, threading devices, knotters, thread tensioners, self-stoppers, bale separators, special yarns. Prevention of fluff accumulation, especially when weaving fibres with high dust formation is possible with the help of automatic fluff blowers. The start and stop buttons ensure that the rewinding machine can be started or stopped from the bobbin side. When refilling a new bobbin rate, a threading device and a knotting device are used for each bobbin tier on both sides of the bobbin. With a six-tier bobbin case, 12 knotters are used. The time of knotting is 5 sec, and the time of moving the knotter is 2 sec. A 600-bobbin creel with 50 vertical racks on one side requires about 6 min. tying the ends of the threads.

Balloon separators are used for the weaving of yarns with high line densities. The lightweight, vertically positioned cylinder separator plates prevent the cylinders from colliding. This increases the yarn winding speeds by 20 to 60 %.

Self-stops monitor the integrity of the reel thread at the bobbin outlet or at the tensioner outlet. If the thread breaks, a very light self-stop pin kinematically closes the electrical circuit and stops the machine. The average reaction time of the self-stop is 6/100 sec. In practice, these values mean that at speeds above 600 m/min, the end of the broken thread travels a distance of 0.6 metres during the reaction time, whereas the complete stopping distance is 3-4 metres.

When weaving crepe (highly twisted) and textured (highly stretchable) yarns, special pleated or plastic sleeves are placed on the surface of the bobbin to prevent the formation of twists when winding the yarn.

An additional bar threader is also used, the angle of bar thread coverage is adjustable depending on the type of material being processed.

Polyolefylene yarns (cut film yarns in the form of ribbons) are rewound by radial rewinding, as axial rewinding causes additional twists of the yarn (ribbon). Due to the large mass of the bobbin (up to 5 kg.) during winding there are large inertia, so it is necessary to provide for braking of the rotating bobbin when reducing the speed of winding (machine stop) and unbraking of the bobbin when increasing the speed of winding (machine start). It is important to prevent the build-up of static electricity by means of charge neutralisers, which are installed on the bobbin or directly on the rewinding machine.

The rewinder cannot do without downtimes associated with bobbin refuelling. The less the downtime, the higher the process efficiency. This is to a large extent facilitated by group filling of threads when changing the bobbin rate (replacement of the thread tying process), the use of bobbins with trolley bobbins, rotary bobbins, magazine bobbins and wedge bobbins. However, the most economical solution is the use of at least two standard creels per spinning machine. While one bobbin is used during the rewinding process, the other bobbin is used for filling. In this case, the creels are additionally equipped with reed holders, into which the threads from the stock are inserted. The used bobbin is then moved aside and the threaded bobbin with reed holder is installed in its place. In this way, the time lost during the bobbin changeover is kept to an absolute minimum. The reduction of the bobbin changeover time, taking into account the further automation of the bobbin changeover, is one of the ways to increase the efficiency of the weaving process.

2.3. Thread tension during the weaving process

The main technological requirement of the process is to create a uniform tension of all warp yarns during shearing.

Excessive yarn tension causes high drafting and reduces the elastic properties of the yarn.

Insufficient thread tension does not ensure the specified winding density, which results in a disturbance of the winding structure and threads of the upper layers crashing into the lower layers of the winding.

Uneven tension of all warp yarns disturbs the cylindricality of the winding on the mending roller or on the warp and leads to the formation of hollows and bulges on the winding surface.

The above-mentioned disadvantages lead to increased breakage in the process of weaving and weaving, to the deterioration of the quality of the fabrics produced.

The bobbins are equipped with thread tensioners in order to obtain the required and uniform tension of all threads during the weaving process.

Thread tensioners are subdivided according to the following characteristics:

- by the way the thread tension is created;
- by the nature of the force that exerts tension on the thread in the tensioner;
- by the presence or absence of automatic adjustment of the output thread tension.

The following methods of creating thread tension are known:

1. By sandwiching it between two surfaces;
2. By enveloping it on a fixed curved surface or multiple surfaces;
3. By braking the roller or disc rotating the thread.

In the first type of tensioner, the thread is straight and clamped between two surfaces (Fig. 14). The tension created by the tensioner is created by friction between the thread and the parts clamping the thread.

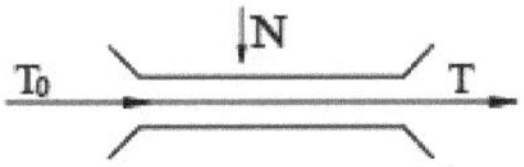

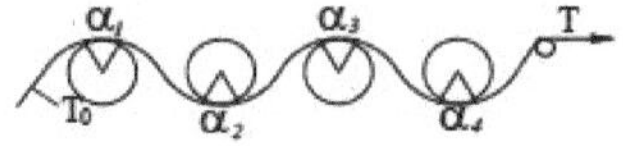

Fig.14. Washer tensioner.

The tension is determined by the following relationship:

$T = To + Nf1 + Nf2$

where: To - thread tension up to the tensioner;

N is the normal pressure on the filament;

f_1 is the coefficient of the top plate;

f_2 - coefficient of friction of the thread against the surface of the bottom plate.

If f1 = f2 then

$T = To + 2Nf H$

The thread tensions are adjusted by changing the normal pressure N.

In tensioners that create the tension of the thread in the second way (Fig. 15) is determined by the well-known Euler's formula

$T = To \exp f (\alpha_1 + \alpha_2 + \alpha + \alpha)_{34}$

where: $\alpha1, \alpha, \alpha, \alpha_{234}$ - the angle of thread coverage of each rod; f - the coefficient of friction of the thread on the surface of the rod; To - the initial tension of the thread. By changing the angle of thread coverage of the rod a, the required technological tension of the thread is achieved.

Fig. 15. Ridge tensioner.

In the third method (Fig. 16), the thread tension is approximated by the expression:

$$T = To + F\,r/R$$

where: R - radius of the idler roller; r - radius of the brake roller; F - braking force of the brake roller.

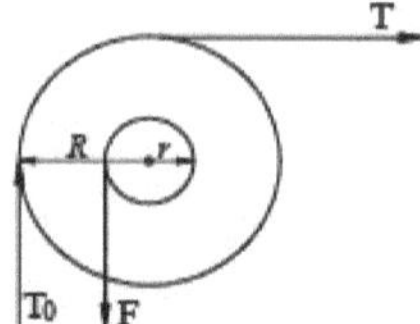

Fig. 16.Disc tensioner.

Proper operation of the tensioner is possible when the friction moment between the thread and the roller is greater than the braking force of the roller.

Otherwise, the thread starts sliding on the roller, the roller stops rotating and acts as a thread brake. In this method, the thread tension varies depending on the force applied to the brake roller.

The nature of the applied force in all tensioning methods can be weight force, spring force, magnetic forces, atmospheric pressure force and initial thread tensioning force. These types of forces can be applied in any type of tensioner. Tensioners with automatic adjustment of the output thread tension are damped and undamped. The damped ones help to damp the oscillation of the moving parts of the tensioner, which are taken out of equilibrium by one or another tension jump.

With the help of a special sensitive organ, which controls the value of the output thread tension, the forces acting on the thread braking element are automatically adjusted (corrected).

Differences in speed during the start-up, shut-down and operation of the rewinding machine, in the positioning of the feeding bundles, in the height and length of the bobbin, in the winding of threads from feeding bundles with different diameters and winding densities increase the tendency for uneven tensioning of running threads. In order to equalise the tension of the yarns, a stepwise application of normal pressure in the tensioners is set both along the height and along the length of the bobbin. Special rollers can also be used at the bobbin outlet, located at an angle or made in the form of a cone, providing a different angle of thread winding of the roll surfaces (Fig. 17).

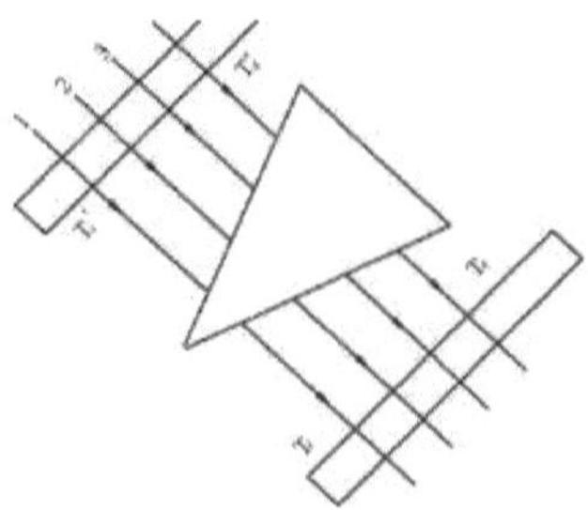

Fig. 17: Cone tensioner.

The thread tension T' *of* the far section of the bobbin is greater than the tension $T4$ of the near section of the bobbin, i.e. $T_1' > T_2' > T_3' > T_4'$ etc. However, according to Euler's formula

$$T = T' \cdot (\exp f\alpha)$$

therefore, the thread tension is the same at the bobbin outlet

$T_1' = T_2' = T_3' = T_4'$ since the larger tension cone T' has a smaller end yarn wrapping angle, and the smaller tension $T.$ has a larger cone yarn wrapping angle. Considering that the thread tension along the height of the bobbin in its middle part is smaller than in the edges of the bobbin, the dimensions (diameter) of the cone towards the edges can be reduced.

It is advisable to install special rollers at the outlet of the tensioner Fig. 18. The thread 1 passes the tensioner (not shown in Fig. 18) guides 2 and a special roller 3, installed with the possibility of movement. Depending on the operating mode of the machine, the roller 3 automatically takes one of three positions.

First - full opening (Fig.18a), the roller 3 is in the extreme left position, when stopping to provide free access to the bobbins during threading.

Second - maximum thread wrapping of the roller (Fig.18b), roller 3 is in the rightmost position, during the start - stop of the machine to provide additional tension of the thread.

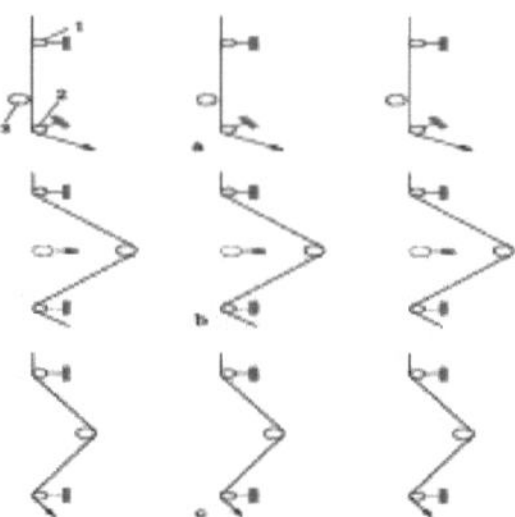

Machine stop period
Start-stop period of the machine
Shear period

Figure 18: Step tensioner.

In this position, the yarn brakes smoothly depending on the weaving speed during the start and stop period, the sagging yarns are partially picked out, which prevents yarn twists.

Thirdly, during the weaving period, the roller 3 (Fig. 18c) takes such a position, in which, as the thread moves away from the front bobbin (along the length of the bobbin) to the back bobbin, the angle of twisting of the thread 1 on the roller 3 decreases. This ensures that the tension is equalised for all the threads running together.

Tensioners with pressure rollers are used for weaving glass and elastomeric yarns, high-strength technical yarns and flower bases that are very sensitive to friction, bending and buckling. Due to the large bending radius, the thread passes through the sting of the retarded rollers, driving them in motion, and damage to the elementary threads is avoided when the threads run around the rollers.

Also yarns with different colours (white, blue, yellow etc.) have different friction coefficients, to some extent yarn tensioners with pressure rollers do not pick up these irregularities.

The automation of the pressure force of the rollers against each other, at different operating modes (start-stop, operation, prolonged stop) of the machine and at different bobbin diameters, ensures uniform tension of the thread during the weaving process.

2.4 Batch weaving

Batch weaving is the most productive and is widely used in weaving for processing all types of yarns, including artificial and synthetic silk, complex and polyester yarns.

Modern batch re-sewing machines must have versatility, high productivity, build up a quality backbone and consider ergonomic design aspects.

Versatility. The batch spinning machines process all types of yarns with linear densities from 7.5 to 170 tex. For yarns with a high linear density there is a balloon stopper in the form of two parallel bars preventing the neighbouring yarns from being caught and installed with the possibility of extending during bobbin change. The possibility of working on the rewinding rollers soft and normal winding. Change of the number of yarns, changeover from one type of yarn to another type of yarn (e.g. from plain to twisted yarn) due to quick re-filling of the yarn group and reduction of guiding elements (rows, reeds, price sticks, etc.).

Performance. Powerful drive, efficient and synchronised braking of all rotating elements of the rewind, presser and measuring rolls. Short time of manual operations, automation of the process of changing rolls on the machine and bobbin stakes, quick elimination of yarn breakage, increase of the shearing speed up to 1200 m/min at a flange spacing of 1000 mm allows to increase the productivity of the rewinding machines. In the case of shearing, the average speed is lower than the calculated speed due to acceleration and stopping of the rewinding roller because of yarn breakage. The difference increases with increasing speed.

The average weaving speed is determined by the following formula:

$$V = \frac{2LV_p}{2L + V_p(t_1 + t_2)(r_o + 1)}$$

where: L - length of thread wound on the mending roller, m; V_p - design (nominal) speed of thread m/min; t1 - roller acceleration time min; t2 - roller stopping time min; ro - number of thread revolutions falling on one roller.

The actual speed, taking into account the time for breakage elimination and the number of thread breaks per needle roller, can be determined as follows Expression

$$V_\phi = \frac{V_{cp}}{(1 + r_o \cdot t_3 / 100)}$$

where: t3 - time of thread break elimination from 0.7 to 1.12 min.

The braking distance S should not exceed 4 metres at speed v_p =1000 m/min, which prevents broken thread ends from running onto the roller.

The stopping time of the spinning roller can be determined

$$t_2 = \frac{S}{V_p}$$

The acceleration time is usually taken depending on the design speed tp = 0.1 - 0.17 (min). Table 8 shows the results of calculations of the average and actual speeds for shearing yarns with a linear density of 15 tex, 60000 metres of shearing length, at a design speed Vp = 1000 m/min, depending on the number of thread breaks.

Table 8.

Calculations of average and actual shearing speeds.

№	Number of breaks per 1 million metres of yarn r	Number of breaks per roll Goh	Stop time of the roller, t_2	Roller acceleration time t_1	Time for break elimination t_3	Average speed, V_{cp}	Actual speed, V_f
	Obr.	Obr.	min.	Min.	min.	m/min	m/min
1	One	36	0,004	0,166	1,12	950	677
2	Two	72	0,004	0,166	1,12	906	502
3	Four	144	0,004	0,166	1,12	830	318
4	six	216	0,004	0,166	1,12	765	224

As can be seen from Table 8, when the thread breakage increases by a factor of six, the actual knotting speed decreases by a factor of three.

Improving the quality of the input bundles, reduces thread breakage and increases machine speeds, thus improving process efficiency. The bobbin rate in the rewinding frame also has a significant effect on the efficiency of the rewinding process. As the bobbin rate increases, the number of bobbin changes decreases, but at the same time the time required to eliminate thread breaks increases. When the bobbin rate is increased up to a certain value, the productivity of the rewinding machine increases and then starts to fall. The following expressions are used to determine the optimum bobbin rate:

for continuous weaving

$$m_{on} = 6000/\sqrt{av*c}$$

discontinuous weaving

$$m_{on} = 1000*\sqrt{32}/\sqrt{avc}$$

where: a is the number of breaks per 10^6 m of single thread; V is the shearing speed m/s;

C - coefficient, which takes into account transitions of the rewinder at elimination of breakage. C = 1.4 + 1.5 - for continuous weaving. C = 0.4 + *0.5 -* for discontinuous weaving.

Modern batch reweaving machines have a high production capacity - one reweaving machine can service 500 weaving machines. When developing new batch machines, the aim is not to increase productivity but to improve the weaving quality.

The quality of the warp can be achieved by winding the same length of threads on the rewinding roller with the same tension of all the threads and with the correct positioning of the threads on the roller. For this purpose: the body of the winding on the rewinding roller should be perfectly cylindrical; the tension of the threads remained unchanged during the entire process of rewinding, regardless of the speed of rewinding, the diameter of the bobbins, their placement on the rewinding frame and other factors affecting the tension of the threads; picking threads through the working bodies should have the correct order and prevent their crossing. Poor quality warp has uneven tension of threads and not the same length. The different lengths lead to sagging or overstretching of the threads on the sander and on the weaving machine. Loose (sagging) threads cause the threads to twist when threading. When these yarns pass to the separating bars of the price field (after the drying reels), the yarns break and the sanding machines are downtime. On the weaving machine, the slack threads stick to other threads and are cut off by the shuttle, rapier or weaver. The overstretched threads are further deformed during the weaving process, which also leads to thread breaks. Uneven (unequal) tension of threads across the width of the warp, causes different processing of warp in the fabric, which leads to visible defects in the fabric and in the subsequent dyeing of the fabric to its different shades, due to different absorption of dye. Besides, at finishing of some assortments of fabrics there can be problems at application of coatings on fabric (glue, rubber, etc.), at gluing of fabrics, at thermofixation. Thus, the quality of the warp obtained during weaving has a great influence on the productivity of the weaving machine and on the quality of the woven and finished fabric.

The batch spinning machine is designed to produce high quality bases:
- accurate thread feeding;
- uniform pressure of the rolling roller against the spinning roller;
- recoil of the rolling roller in the event of a thread break;
- maintaining a constant warp tension;
- accurate measurement of the specified base length;
- oscillatory movements of the dividing row.

Careful filament feeding, thanks to the minimum distance between the separating row, the guide roller and the winding point on the rewinding roller, ensures high filament feeding accuracy when the filament is fed onto the rewinding roller. The indirect pressure system (Fig. 19) ensures uniform pressure of the rolling roller on the rewinding roller. When working up the rewinding roller 1, the increase of the winding diameter 2 leads to the movement to the left of the rewinding roller 3, loaded by the pressing system 4 (Fig. 19).

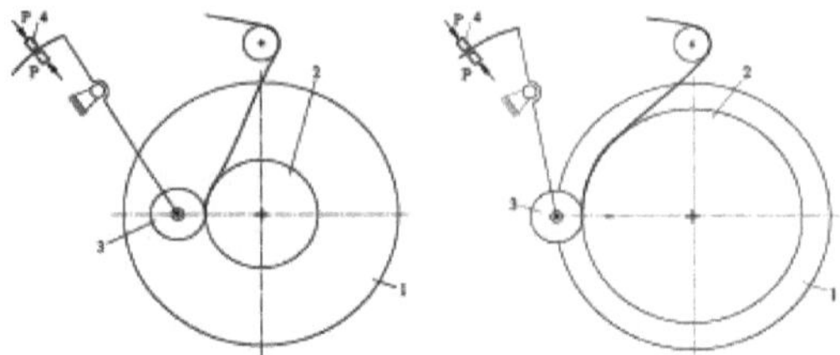

Fig. 19: Indirect press system of the rolling roller to the spinning roller.

This indirect pressure system favours a perfectly cylindrical winding on the spinning roller.

The recoil of the rolling roller during the braking period prevents friction between the rolling roller and the warp yarns wound on the rewinding roller. The indirect press system does not require any additional devices, as the recoil of the rolling roller is forced by kinetic energy. When the machine is started, the roll is automatically returned to the working position. The constant yarn speed keeps the yarn tension at the same level.

The yarn tension fluctuations during machine start-up and shut-down are minimised due to the short acceleration and deceleration times of the mending roller. The force interlocking of the weaving roller with the mending roller and the measurement of the warp length by the weaving roller ensures a more accurate warp length measurement than the warp length measurement on the guide roller.

The bars of the separating rows are made with steps corresponding to the number of tiers in the bobbin height to facilitate thread picking into the rows.

The horizontal movement of the separating row distributes the threads evenly over the winding surface of the spinning roller.

The ergonomic aspects of the design ensure: easier labour for the operating personnel; optimum protection for the personnel; well-designed and conveniently located controls. The convenient working height of the creel, the minimum distance between the creel and the rewinding machine, the stepped design of the sliding row rods, the reduced fatigue of the operating personnel allow to reduce the time for the elimination of thread breaks from 1.12 min to 0.6 min. Example: the number of threads in the base 4200 threads, the rate of bobbins 600 threads, thread breakage on the needle roller 36 breaks. Time for elimination of thread breakage:

option 1 - 36 - 1.12 = 40.32 min.

variant 2 - 36 - 0.6 = 21.6 min.

Saves time in eliminating breakage:

per roller 40.32 - 21.6 = 18.72 min.

for a batch of rollers (40.3 - 21.6) - 7 = 131 min. or 2.2 hours.

In 2.2 hours, an additional 1.7 spinning rolls can be worked on.

The most widespread different modifications of partion re-sewing machines were made by Barber-Kohlmann, Schlafhorst, Beninger.

2.5 Ribbon weaving

Ribbon weaving consists of two operations: the successive winding of the ribbons on the reel (winding) and then simultaneously all the ribbons are wound on the weaving head (lenoing). The presence of these operations significantly reduces the productivity of the process and creates uneven yarn tension.

Ribbon weaving significantly reduces waste and makes it possible to produce finished weaving warps with a large number of warp counts. Ribbon weaving is suitable for processing expensive raw materials, warps with a large number of yarns, warps with a large number of yarns with a complex colour ratio in warp lengths up to 2500 m, as well as for setting up production facilities that process yarns or yarns with high physical and mechanical properties twisted in small volumes.

The section of the sliver on the mending drum can have: a rectangle shape, where pins, shackles, shields prevent the edge threads of the sliver from falling off; a parallelogram shape, where the base of the first sliver is placed on the drum cone, and the second on the cone of the first sliver. The correct sliver structure is obtained by adjusting the drum cone angle and caliper feed (movement) depending on the thickness of the threads. Incorrect adjustment will lead to distortion of the sliver cross-sectional shape, and as a consequence to changes in the length and tension of individual threads.

For spinning machines with a variable taper angle on the drum and a constant slide movement, it is characteristic that the taper angle of the drum increases with decreasing linear yarn density, i.e. the finer the yarn, the greater the taper angle of the drum.

For spinning machines with a constant drum cone angle and variable slide feed, it is characteristic that the value of the slide movement increases with increasing yarn line density, i.e. the thicker the yarn, the greater the slide movement.

Hence, in order to obtain a good winding structure and a longer sliver length on the drum, yarns with a high linear density are run on rewinding machines with a constant drum cone angle and yarns with a low linear density on rewinding machines with a variable drum cone angle. There are two ways of regulating the caliper movement: manual - on the basis of the control data of the weaving process, the experience of the operating personnel, using monograms and tabular material, etc.; automatic - on the basis of recording the position of the threads during weaving and automating the caliper movement process.

Let's calculate the value of the slide movement when threads are threaded onto the drum. Define the cross-sectional area of the belt (Fig. 20):

$$S = a \cdot b = a \cdot H \cdot tg\alpha$$

where: a - sliver width, cm; b - thickness of the sliver winding, cm; H - feed rate of the caliper when the whole sliver is wound, cm; a - angle of the sliver taper.
Sliver winding capacity

$$V = S \cdot \pi \cdot D = a \cdot H \cdot \pi \cdot D \cdot tg\alpha$$

where: D - value of average winding diameter, cm

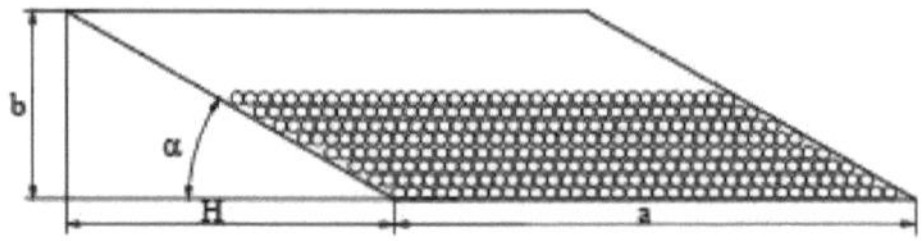

Figure 20: Cross-sectional area of the ribbon.

Tape winding weight:

$$G = V \cdot \gamma = a \cdot H \cdot \pi \cdot D \cdot \gamma \cdot tg\ \alpha$$

where: γ - winding density gr./cm
Weight of one coil of tape

$$g = \pi \cdot D \cdot T\ /\ 1000$$

where: T - linear yarn density, tex.
Total number of yarn turns in the entire sliver.

$$K = G / g = 1000 \cdot a \cdot H \cdot \pi \cdot \gamma \cdot tg\alpha / \pi \cdot D \cdot T = 1000 \cdot a \cdot H \cdot \gamma \cdot tg\alpha / T$$

On the other hand, the number of yarn turns in the sliver

$$K = n \cdot a \cdot p$$

where: n is the number of revolutions of the drum during the thread weaving time;
a - ribbon width, cm; p - ribbon density (number of threads per 1 cm).
Equating equal values we obtain

$$n \cdot a \cdot p = 1000 \cdot a \cdot H \cdot \gamma \cdot tg\ \alpha / T$$
$$p = 1000 \cdot H \cdot \gamma \cdot tg\alpha / n \cdot T$$

Total movement of the slide when the entire sliver is being worn down

$$H = p \cdot n$$

where: n is the movement of the slide per one revolution of the drum.
Let's substitute the obtained expression into the previous one and get

$$p = 1000 \cdot n \cdot \gamma \cdot tg\alpha / T$$

Determine the movement of the slide for one revolution of the drum

$$n = p \cdot T / 1000 \cdot \gamma \cdot tg\alpha$$

The amount of movement of the slide depends on the thickness of the yarns, the density of the yarns in the sliver, the specific density of the yarn winding on the drum and the angle of the drum taper.

Manual adjustment of the caliper feed rate requires certain skills of the operator to adjust the caliper feed rate during the winding process. Otherwise, incorrect increase or decrease of the slide feed, inaccurate setting of the joints between the tapes by the slide leads to non-cylindrical winding surface on the drum, i.e. protrusions or depressions between the subsequent tapes at the drum cone and the first tape.

Automatic control of the slide feed ensures cylindricality of the winding on the drum by measuring constantly or periodically the thickness of the wound warp on the drum and the results of the measurements are fed to the computer, where they are compared and a command to move the slide is issued.

Measurement of the winding thickness on the drum can be carried out by means of a sealing roller in contact with the warp winding.

The compaction roller enables low yarn tension, yarn conservation and a compact (tight) winding. Pressing the yarns with a certain force during winding has an levelling effect, i.e. the same length of all yarns across the filling width of the spinning reel.

In addition to the same length, the threads must have uniform (equal) tension across the entire filling width for the entire period of winding on the reel.

However, deviations in thread tension can be caused by changes in working speed, reduction in bobbin diameter, changes in the coefficient of friction in tensioners and thread guides, changes in thread tension due to batch changes within the same warp.

Thus, when the threads are placed on the spinning reel, the thread tension is influenced by factors not controlled by the thread tensioner.

Measuring the tension of the threads at the point of running into the drum winding body and transmitting the signal to a microcomputer, where it is compared with a predetermined value of thread tension and issuing a command for automatic regulation of the tensioner, i.e. maintaining at the same level the preset tension of all single threads solves this problem.

The slivers are lenoed from the spinning reels to the weaving head under a defined yarn tension.

Excessive yarn tension impairs the physical and mechanical properties of the yarn, causes the upper yarn layers to cut into the lower yarn layers and reduces the leno speed of the yarns.

In order to produce quality warps, the warp is subjected to a longitudinal reciprocating movement which forms a cross winding on the warp. In addition, an additional roller is pressed against the warp winding with a certain amount of force to compact the winding while minimising warp tension.

In some cases it is of interest to determine the relationship between the capacity of the spinning drum and the maximum winding diameter on the reel. The comparison is made on the basis of the winding volumes on the drum and on the reel. The winding volume on the drum is determined by the formula

$$V_6 = \pi \cdot n \cdot S \cdot l \cdot \sin\theta (d + l \cdot \sin\theta)$$

where: n-number of belts; S-movement of the slide, cm; *l-length of* the taper on the drum, cm; d-diameter of the reheating drum, cm; θ-angle of the drum taper, deg. Yarn volume per skein by formula

$$V_{\text{н}} = \frac{\pi \cdot H}{4}(D_{\text{н}} - d_{cm})$$

where: H-arrangement of winding flanges, cm; DH - winding diameter on the winding, cm;

dcm - trunk diameter of the navoi, cm.

The resulting $V_6 = v_H$ volumes are equated

$$\pi \cdot n \cdot S \cdot l \cdot \sin\theta (d + l \cdot \sin\theta) = \pi \cdot H / 4 \left[\left(D_H^2 - d_{cm}^2 \right) \right]$$

The obtained expression, solving with respect to the winding diameter on the head, we obtain $D_H = \sqrt{4 \cdot n \cdot S \cdot l \cdot \sin\theta (d + l \cdot \sin\theta) / H + d_{cm}^2}$

Example; $n = 10, S = 17{,}8cm., l = 85cm., \theta = 13.6град., d = 75{,}5cm., d_{cm} = 18cm., H = 178cm.$

DH=89.2cm

In this way, the specified yarn volume is wound from the spinning drum onto the weaving warp with a winding diameter of 90 cm. Modern spinning machines are equipped with a warp flange diameter of up to 1250 mm.

The productivity of draw frames depends on the bobbin rate, the number of ribbons, the number and type of price cords, the thread breakage, and the machine design (removable or non-removable drum).

Increasing the bobbin rate with a corresponding reduction in the number of tapes does not give the desired increase in machine productivity.

The value of the optimal rate is determined by the following expression

$$m_{on} = 2000\sqrt{b / a \cdot c} \cdot \sqrt{1/V + t/L}$$

where: *b - number of* bobbins in the vertical row; *a - number of* breaks on 10^6 m of single thread; *c* - coefficient determining the time spent on transitions, between two neighbouring rows of bobbins; *V* - speed of shearing, m/sec.; *t* - downtime of machines in the process of re-filling the tape and laying prices in the process of shearing one tape, sec.; *L* - length of shearing, m.

Knitting time per warp

$$t_c = LK_s / V_c \eta_c$$

where: *L-length of* shearing, equal to the warp *length*, m; *Cl-number of* tapes; vc-speed of shearing, m/min; nc-coefficient of useful time of shearing.

Twisting time $t_n = L / V_n \cdot \eta_n$

where: η_n - coefficient of useful lacing time; V_n - velocity of lacing, m/min.

Total manufacturing time of the base: $t0 = tc + tn$

Process throughput in terms of quantity of material processed in kg/h.

$A_1 = A \cdot n_0 \cdot T \cdot 10^{-6}$ кг/ч

where: T - linear yarn density, tex; n_0 - number of yarns in the warp.

Belt rewinding machines are produced by Beninger, Hakoba and Textima.

3. SANDING OF BASES
3.1 Schlichting materials

Schlichting is an operation of applying a solution of schlichte to the thread, which after drying forms a protective film of colloidal, adhesive, binding and film-forming substance.

The ground yarn must be able to withstand vibration, friction, impact, stretching, electrostatic effects during the weaving process and ensure smooth operation of the weaving machine to produce high quality fabrics.

The slurry consists of adhesives (starch, CMC, PVA, etc.) and auxiliary substances (softeners, antiseptics, etc.).

Flaking agents are water-soluble (PVA, PAAM), which can be directly washed out of the fabric with water, and water-soluble (starch, CMC, etc.), which are first converted into a water-soluble state and then washed out of the fabric. The useful properties of the slurry are completed with the production of woven fabric.

The process of desliming the fabric (removing the slurry) is appropriate with water-soluble materials, as they are easily removed and can be reused after regeneration, thus significantly reducing wastewater pollution.

A minimum slurry composition should be aimed for:
- for cotton yarns - CMC, starch, wetting agent or CMC and wetting agent;
- for woollen yarns - CMC, wetting agent or PAAM and wetting agent;
- for viscose yarn - starch, CMC;
- for viscose and acetate yarns - PVS or PAAM;
- for blends of polyester yarn with viscose fibre - starch and CMC or starch and PVA.

The quantitative and qualitative characteristics of the sander depend on the type of yarn, yarn line density, fabric weave, type of sander and weaving machine.

The interrelation and optimal combination of factors - adhesion, adhesion, adhesion of the slurry to the fibres, penetration of the slurry into the yarn, uniformity of its distribution on the yarn, strength and flexibility of the slurry films - with controlled parameters (slurry composition, concentration, viscosity, slurry temperature, drying speed and temperature) ensure high efficiency of the slitting process and quality of the obtained semi-finished products.

The slurry parameters are kept constant during the cooking process. In order to ensure a given slurry concentration and viscosity, the dry matter weight and slurry volume must be matched and dosing devices, means and instruments for measuring and controlling the viscosity, concentration and volume of the slurry must be used.

Nowadays, the automatic system of slurry preparation is widely used (Fig.21). When selecting the slurry composition, the operator transmits by digital code the task to microprocessor (computer) 1, which determines the correct set of all controlled parameters - temperature, pressure, speed, viscosity, concentration, slurry volume and specification of slurry components. All information is given on the display for the operator to see.

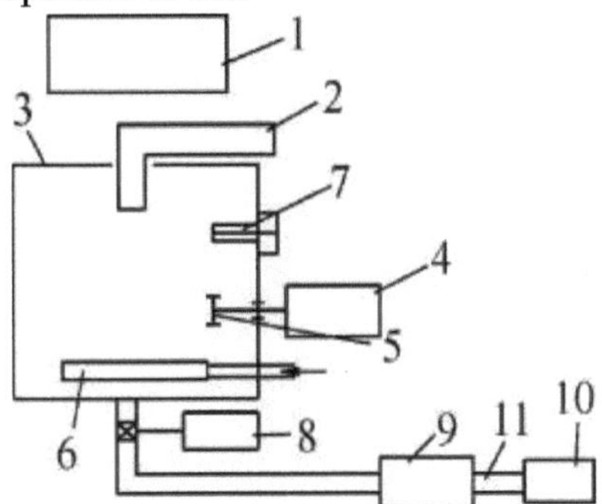

Figure 21: Automatic slurry preparation system.

Microprocessor 1 gives the command: to pour water 2 through the pipeline into the tank 3 of the set volume; to switch on the electric motor 4 stirrers 5; to load each suspended component into the tank 3; to supply steam 6 to the coil through the steam line. When the set temperature is reached, the control system 7 keeps it constant during the cooking process.

At the end of cooking, the slurry is transported by pump 8 to tank 9 for storing the slurry and maintaining the temperature of the slurry at a certain level. When a signal is received from the glue trough 10, the slurry is transported to the sanding machine via the slurry line 11.

In the final stage of the grinding process, the microprocessor determines the quality of the slurry and compares it with the data stored in the programme.

The reuse of slurry removed from the fabric during the desanding process solves the problem of wastewater treatment and the cost of slurry materials.

Raschlichting of fabrics allows to remove from the fabrics up to 95% of the slurry applied on it in the form of a solution with a low concentration of up to 30 g/litre, and for sanding it is necessary to have a concentration of 80-120 g/litre.

Evaporation, precipitation and ultrafiltration are used to increase the slurry concentration. However, these are costly and do not produce a slurry with well reproduced properties.

For regeneration of hygroscopic, rapidly swelling and water-soluble slurry the water-soluble slurry removal apparatus of Beninger company is used (Fig.22).

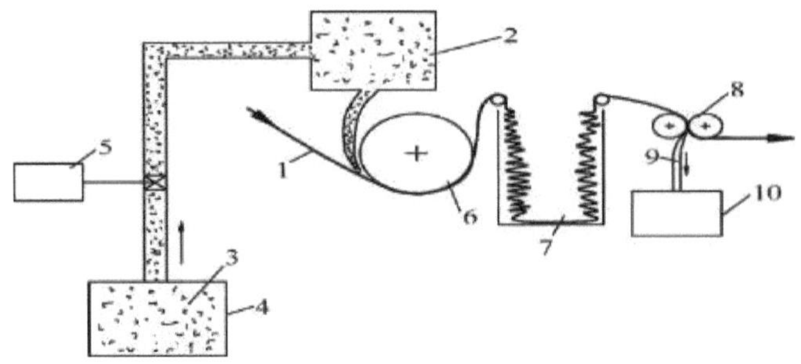

Fig. 22:
Beninger water-soluble slurry remover

Fabric 1 is treated in the area of slurry 2 with the solution of slurry 3 with the concentration of 40-60 g/litre, supplied from the tank 4 by means of the pump-dator 5. Further, the fabric envelopes the shaft 6 and enters the accumulator 7. After swelling and dissolution of the slurry, the fabric is squeezed in the squeezing device 8 and through the pipeline 9 the solution with a concentration of 80-120 g./l goes to the tank 10. The fabric, when moving into the area of application of the slurry, causes foaming of the solution, due to the large contribution of air flow of the fabric into the solution, resulting in a decrease in the area of contact between the fabric and the solution. At the speed of fabric movement 50 m/min and fabric weight 200 g/m^2 , the amount of air introduced by the fabric into the solution is 15 litres/min. Feeding the fabric and solution (slurry) under the enveloped shaft 6, squeezes out the air from the fabric into the atmosphere with viscous solution, the amount of fed slurry (solution) is 150% of the fabric mass. It is expedient to use the apparatus in a set with washing and drying machines.

3.2 Application of the slurry.

There are three ways of applying the slurry to the thread:
-   sanding with soluble, emulsified and dispersed substances with water (classical);
-   wet sanding with soluble, emulsified and dispersed substances with solvents (perchloroethylene, perchloroethane);
-   dry sanding with wax-like solids or molten substances.

The classic slurry application is carried out by using a highly concentrated slurry, a high pressure force on the squeegee rollers and a foamed slurry. Highly concentrated slurry reduces the energy required for drying the substrate. The degree of absorption of the solution by the yarn is not due to immersion of the warp 1 in the slurry 2 (Fig. 23), but due to feeding the warp with solution in the sting of squeezing rollers 3 and 4. Supply of liquor into the sting of squeezing rollers is carried out by the lower roller 4, and the amount of slurry applied to the warp is carried out by the roller 5.

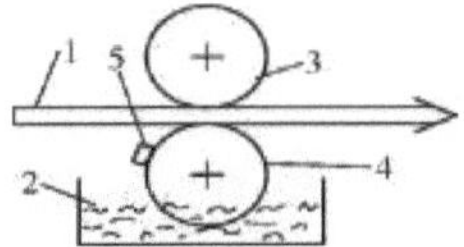

Figure 23.Glue trough.

In order to reduce pressure in the stinger and optimal dosage of grout on the base, the upper shaft has a rubber coating and the lower shaft has an engraved surface. Excess liquor is removed by the rake 5 from the protruding surfaces of the shaft 4, and the liquor remaining in the recesses of the shaft goes into the stinger of the shafts. The rubber of the upper roller presses the warp into the recesses of the surface of the roller 4 and the slurry is applied on one side of the warp threads.

Disadvantages of this method:
- when changing the range of fabrics, difficulties in replacing the engraved roll;
- depending on the amount of glue, the rollers have a different structure and depth of engraving;
- contamination of the engraving recesses or mechanical damage to the roll surface will change the degree of filament finishing.

The main advantage of the method is that the degree of filament finishes does not depend on the speed of sanding.

Conventional slitting uses roll spinning from 15 to 60 kN and processes low to medium specific gravity yarns in a low viscosity slurry.

Squeezing the warp on rollers with high pressing force (up to 100 kN) during sanding reduces the moisture content of the yarns at the outlet of the sanding trough and reduces the energy consumption for drying the warp.

Consequently, the high slub concentration and the high spinning force of the rollers reduce the consumption of sanding materials and the energy required for warp drying, increase the sanding speed and the processing capacity of the warp in weaving. These developments have been applied to the West Point spinning machines.

Foam sanding has the same advantages as sanding with high squeezing force and high concentration of sand.

Foam is produced by mechanical mixing of the solution: in the presence of air; by blowing in compressed gas; by heating high quality emulsions consisting of foaming liquids.

The composition of the foam solution has a solvent (water), inorganic solvents, adhesives, foaming agent, additives that regulate the viscosity or stability of the foam.

Foam is applied by knife or roller rake in horizontal or vertical plusovka. During the movement of the base 1, the foam 2 is evenly distributed by the knife rake 3 (can be a roller rake) and then squeezed by the rollers 4 (Fig. 24). Suction devices 5 are installed in the area of squeezing rollers - knife rake.

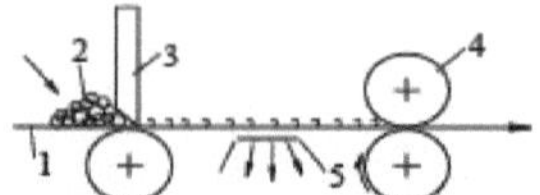

Fig. 24.Horizontal plussing.

In the horizontal plusovka (Fig. 25), the warp 1 is passed through the first squeezing device 2 for uniform distribution of foam 3 over the entire web of warp yarns (the gap between the squeezing rollers is 3-5 mm). The second squeezing device 4 presses the foam into the yarns with force, this leads to the destruction of the foam and its penetration into the surface layers of the warp yarns.

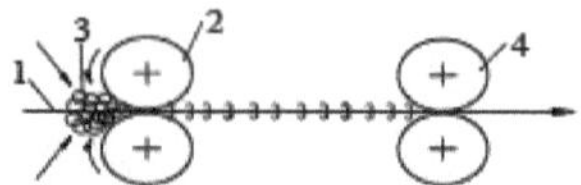

Fig. 25: Horizontal plussing.

In the vertical plough (Fig. 26), the foam is applied on both sides and squeezed in the plough rollers. During squeezing, the foam disintegrates, viscosity decreases, which improves the distribution and penetration of the mortar into the substrate.

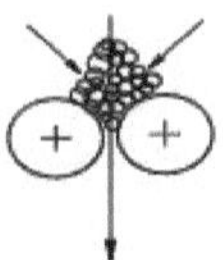

Fig.26.Vertical plussing.

The adhesive is controlled by the concentration of the slurry, the filling of the solution with air and the volume of foam.

The speed of sanding has a great influence on the glue. Since the sander is running slowly and stops are frequent, attention must be paid to the even application of the sander on the warp yarns, for this purpose it is advisable to use a horizontal tedder.

The disadvantages of slubbing in foam are the difficulty of ensuring foam stability before spinning and the destruction (breakdown) of the foam at the end of spinning, as well as achieving a viscosity depending on the yarn type and warp density.

Karl Mayer also uses a special sanding bath for sanding, which makes it possible to dispense with the classical wetting by placing in the pre- and main sanding solution and instead use an economical spray application system. The combination of spraying and subsequent squeezing rollers during the sanding process creates three highly turbulent application zones with intensive flow, in which a homogeneous film of substance is formed on a group of filaments in the form of a web, reducing the consumption of spray liquid. The compact design of the material guiding main rollers allows the warp to be processed over a significantly wider range of warp yarn linear densities. In each case, the yarns are guided without crossing and untwisted. For finishing effects, high concentration and viscosity slurries as well as pastes and foams can be used.

Wet sanding in solvent media eliminates uneven distribution of the sand due to the low surface tension of the solvent and its ability to fully wet the fibres.

Methylene chloride, trichloroethylene and perchloroethylene are used as solvents. The solvents used must have the following properties:
- must not be toxic or flammable;
- the cost must be reasonable;
- must not be volatile (evaporate quickly);
- possibility of low regeneration costs (reuse);
- must not cause corrosion of the machine components.

The slurry is cold applied to the warp yarns in a sanding bath. After spinning, the yarns pass through a drying zone where rotating price bars separate the warp into groups of lower density, a hot steam stream treats the yarns and most of the solvent is removed from the yarn surface.

The warp is then transferred to the drying drums and the solvent is finally removed from the yarns. The evaporated solvent is regenerated and goes into the solvent collector. After final drying, the warp is transferred to the dividing area and wound onto the weaving warp.

The yarn is smooth and has no protruding fibre ends due to the effective separation of the warp yarns in the wet and dried state in the drying and dividing zones.

The sanding materials used should be readily soluble in solvent and water, and easily removed from the fabrics when desanding with a water medium.

Wet sanding in solvent media has the following advantages:

- reduces the energy required for drying the base by up to 90%;
- compactness and less space requirement for sanding machines;
- simplifies drying and separation of warp yarns;
- improves the processing capacity of yarn in weaving;
- complete solvent recovery, prevents contamination of wastewater during fabric brushing;
- up to 80 % reuse of the slurry (after regeneration) reduces the costs of sanding materials.

The slight loss of yarn strength when wetting the yarn with solvent can be attributed to the disadvantages of solvent sanding.

Dry sanding (melt sanding) is carried out by applying melted active sand to the warp yarns during the weaving process at a speed of 500-600 m/min.

The applicator roller for applying the melt is positioned between the creel and the winding device (spinning drum or spinning roller).

The applicator roller is heated to 200^0 C and rotated in the direction of the thread movement at a frequency of 10 min^{-1} . As each filament passes through its troughs at high speed, the molten slurry is applied to the filaments with simultaneous smoothing of the protruding fibre tips. After leaving the roller, the warp yarns are exposed to the cooled air and the slurry hardens, strengthening the fibres.

The adhesion value is influenced by the rotational speed of the applicator roll and its temperature, the contact time of the yarn with the applicator roll, the yarn speed and the feed of the filling material to the applicator roll. By adjusting any of the parameters, the optimum mode of spinning of different yarns is achieved.

The high temperature of the applicator roll has no effect on the yarn properties, as the yarn is in contact with the roll for a fraction of a second.

For example, at a shearing speed of 550 m/min and a contact arc of 0.076 m, the yarn contact time with the roller is 0.008 sec. The yarn heats up due to contact with the molten slub and then cools down rapidly.

Advantages of this method:
- reduces yarn linting, increases the processing capacity of warp yarns in weaving;
- high sanding speed;
- 80% energy saving by eliminating drying;
- simplification of slurry preparation;
- combining the technology with existing yarn winding processes;
- The fabrics are dyed according to the known technology without the use of solvents.

Dry sanding increases the abrasion resistance of the yarn, but does not increase the yarn strength.

The disadvantages include - finding new materials for the slurry and developing new means of melt application to increase the strength of the yarns.

3.3. Winding the main threads

The warp yarns are wound up in the following variants: from the bobbin, from the spinning rolls, from the filling or from the spinning drum. The choice of option depends on the changeover time of the feeding forging.

The arrangement of the spinning rollers on the stands of the sanding machines can be single-deck sequential or parallel, double-deck sequential or parallel.

The single tier sequential rack is used when spinning high and medium yarn counts in two baths.

The single tier parallel rack allows separate warp feeding from each spinning roll and is used for processing low yarn count yarns.

The two-tier parallel racks ensure a straight path of the warp from each roll and are used for monofilament sanding.

The two-tiered consecutive racks provide easy access to the reamer rollers, which are mounted vertically in three rows, so that between two adjacent frames there is a passage where it is convenient for the worker to monitor the threads coming off the rollers and correct defects.

The disadvantage of double-deck racks is the difficulty of filling the reamer rolls.

The time required for filling new filling rollers accounts for 80% of the total downtime of a spinning machine.

Reducing downtime is possible by automatically replacing the used racks with spare racks pre-filled with spinning rollers.

Replacement is carried out by laterally shifting the racks along the rails or by extending the reeders from behind the racks.

A typical stand can accommodate 16 100 cm flange diameter re-shafts with a mass of 800 kg, which corresponds to a total rotating mass of 13 tonnes. A braking torque of 17 Nm is required to stop the filling roll at a sanding speed of 100 m/min.

For braking of the rewind roller, a rope or belt brake is used, where the braking force is transmitted to the roller by a loaded rope or belt; a polyethylene support, which transmits the braking force to the neck of the rewind roller in proportion to the weight of the winding on the roller.

Winding of the yarns from the spinning rollers under the following condition when the winding torque is greater than the braking torque (Fig. 27).

$$M_{CM} \geq M_{T} \quad \text{or} \quad FR \geq (F_1 - F_2)\, r$$

where T - warp tension; R - warp radius; F_1 and F_2 - tension of the running down and running up branches of the belt; r - braking radius.

According to Euler's formula

$$F_1 = F_2 (\exp f\alpha)$$

where: f - coefficient of friction of the rope (belt) on the brake disc; a - angle of circumference of the brake disc by the rope (belt).

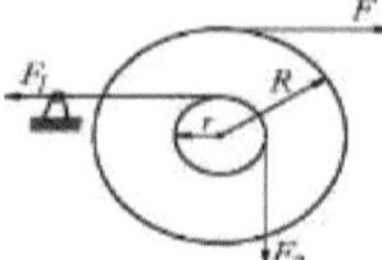

Fig.27. Band brake.

It follows that:
$$F = \frac{F_2 \cdot r \cdot \left(e^{f\alpha} - 1\right)}{R}$$

As can be seen, the warp tension increases when winding off the rollers due to the reduction of the winding radius. It is therefore necessary to manually adjust the rope (sliver) tension, i.e. reduce it.

If the ratio F_2 / R is constant $F_2 / R = const$ over the entire winding operation period, the warp tension will be stable.

This is possible by linking the change of the winding radius to the sliver tension, by using corrective means of warp tension. The braking of the spinning roller by the support friction (Fig. 28) will have the following form

$$F_{TP} = N \cdot f$$

where: N-normal roller pressure on the support; f-coefficient of friction of the shaft journal against the support.

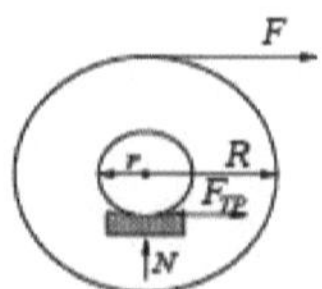

Figure 28: Support friction brake.

On the other hand, the normal pressure depends on the mass of the winding roll

$$N = m - g$$

where: g - acceleration of free fall; m - mass of the spinning roller.

Let's substitute the expression into the previous one and we have

$$F_{mp} = m \cdot g \cdot f$$

Equilibrium equation of the system according to Dalembert's principle

$$FR = F_{mp} \cdot r$$

Hence

$$F = \frac{F_{mp} \cdot r}{R}$$

After substitutions we have

$$F = \frac{m \cdot g \cdot f \cdot r}{R}$$

As you can see with the decrease of the winding radius on the roller synchronously decreases and the mass of the rewinding roller, that is, the ratio m/R value is constant, so the warp tension for the entire period of operation of the rewinding roller is constant.

The total filling tension of the warp yarns can be varied by selecting the required diameter (r) of the brake disc and the corresponding selection of the rubbing surfaces (by changing the friction coefficient f).

During the acceleration period of the spinning roller (Fig. 29), different loads are applied to the warp yarns.

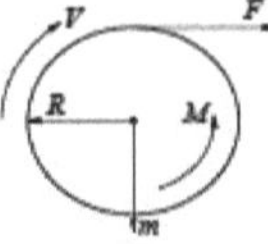

Fig. 29.Loads on the filaments during roll acceleration.

According to the laws of dynamics, the momentum is equal to

$$M = \frac{I \cdot \omega}{t}$$

where: I-moment of inertia of the roller; ω-angular velocity of the roller; t-time of acceleration of the roller.

Since ω= V / R

$$I = (m\text{-}R^2) / 2$$

where: V is the speed of thread winding; m is the mass of the needle roller; R is the radius of winding of the needle roller.

Substituting the obtained expression into the previous one, we have

$$M = \frac{m \cdot R \cdot V}{2t}$$

According to the equilibrium of moments

$$M = \frac{m \cdot R \cdot V}{2t} = F \cdot R$$

Hence the tension of the threads is

$$F = \frac{m \cdot V}{2t}$$

An analysis of the formula shows that high speeds and high mass of the spinning rolls result in high acceleration forces, and the acceleration time must be long in order to maintain the thread tension within the specified limits. The acceleration time of the roller to the full working speed of the machine is adjusted depending on the sensitivity of the yarns to deformation (stretching).

To reduce the draw and to ensure consistency of warp tension use: controlled brakes, which when reducing speed or stopping the machine additionally produce braking of rollers, which eliminates their inertial movement; adjustable brakes, which with increasing warp tension reduces the braking of the spinning rollers.

3.4.Stitching and drying equipment

Modern sanding machines allow many variations in the passage of the threads in the sanding unit to obtain optimum absorption of the sand by the warp threads. The warp can pass through one or two impregnation baths and have one or more dip and squeeze rollers.

The sanding devices are characterised by: direct or indirect heating of the sand and its regulation to maintain the set temperature of the sand to avoid deviations in the viscosity of the sand; regulation of the level of the sand; complete insulation of the impregnation bath to reduce heat losses, vapour formation above the bath and to eliminate the formation of films on the sand; continuous circulation of the sand; adjustable pressure of the squeezing rollers from 0 to 100 kN with automatic control on silent running; manufacture of stainless steel parts in contact with the sand.

applying slurry to the substrate and adjusting the amount of adhesive.

The absorption of the base of the slurry depends on the length of time the filaments are in the sanding bath, determined by the position in height and width of the bath of the plunging (dipping) rollers.

The press rollers press the slurry into the warp yarns, squeeze the excess slurry from the warp, and ensure minimum warp moisture.

Incorrectly selected press roll pressure on the machine at low and working speeds, uneven press roll widths result in over or under gluing of the warp length and width, which is especially noticeable when processing low yarn counts and yarns with a high yarn count.

The specific squeezing force is determined by the formula

$$P = \frac{F}{l \cdot H}$$

where: F - squeezing force; N - warp width in the squeezing rolls; l - length of the clamped warp section, depending on the elasticity of the roll cover.

As can be seen from the formula for a constant shaft load, the magnitude of the specific squeezing force increases as the length of the clamped warp section decreases.

The flexible and equally pressed press rollers, due to the bending of the rollers, contribute to a lower warp press in the web edges than in the middle. The warp has therefore a higher degree of polishing at the edges than in the middle, which should have a favourable effect on the processing properties of the yarn during weaving.

It is advisable to divide the warp yarns into groups after the slubbing process before entering the drying section of the machine. This prevents the warp threads from sticking together and is possible with: the use of several sanding baths (especially in the case of a large number of threads) with drying in the pre-drying zone 1 and 2 of the warp threads 3 and the joining of all threads in the final drying zone 4 (Fig. 30); wet separation of the warp threads 1 with the help of driven slowly rotating polished thread separation bars 2 between the squeezing rollers 3 and drying drums 4 (Fig. 31).

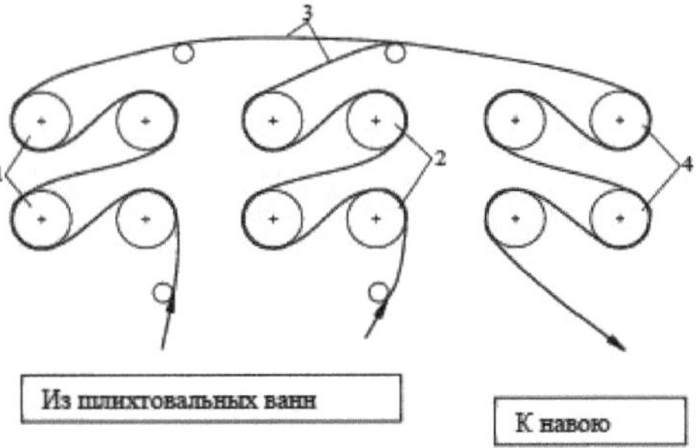

From the sanding baths To the navoi
Fig. 31: Thread separator.

Drying of polished yarns is carried out by contact, convection and radiation methods.

At contact of threads 1 with the hot surface of drum 2 (Fig. 32), the drying process takes place from one side, vapour escapes through the free surface of threads and liquid with particles of slurry moves to the opposite side of the contact surface. Passage of yarns through several drums changes the contact points and leads to yarn diameter deformation.

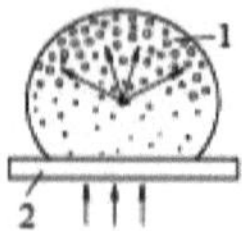

Figure 32. Contact drying.

In convection drying (Fig. 33), the surface of the yarn is dried first. The liquid and slurry particles move to the periphery of the yarn, resulting in the formation of a uniformly distributed slurry film on the yarn surface. With further drying, the movement of moisture to the periphery becomes more difficult and the movement of the yarns in the wet state with the free movement of the warp causes large deformations.

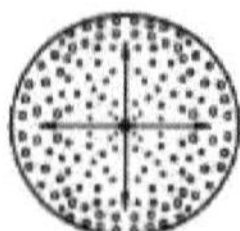

Figure 33. Convection drying.

In radiant beam drying (Figure 34), the liquid moves towards the periphery and the slurry particles towards the centre of the filament.

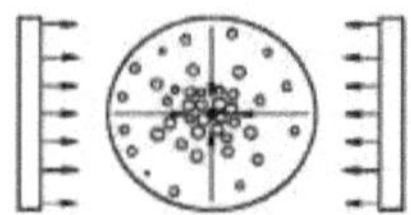

Figure 34: Radiation drying.

It is therefore to be expected (especially for yarns with high line density) that the yarn centre is over-dried and the yarn periphery is under-dried (does not reach the optimum moisture content). Due to the high energy costs, radiation drying is used in combination with other drying methods.

Let's consider the drying process of hygroscopic material presented in Fig. 35. On the first section 1 of the curve up to the point A there is evaporation of liquid from the microcapillaries of the thread surface, from inside the thread the liquid is continuously supplied under the action of capillary forces. The drying speed V_A is constant.

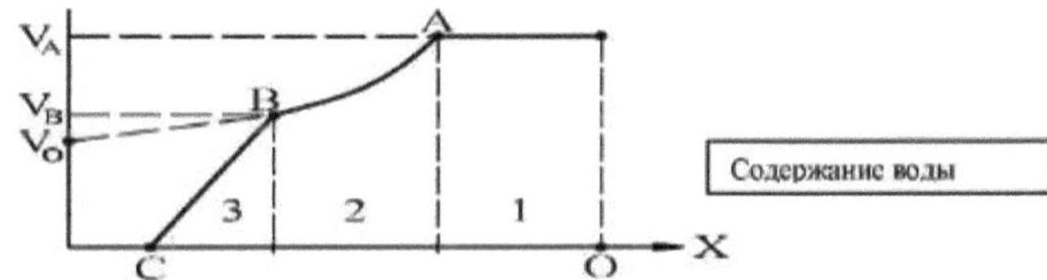

Figure 35. Drying process of hygroscopic material.

In section 2 between points A and B, the drying rate V_A tends to the apparent final drying rate V_B . When the moisture content of the filaments decreases, vapour formation takes place in the microcapillaries, so there is a decrease in vapour pressure.

As drying continues, the differential vapour pressure decreases continuously and the drying speed vB at point B falls linearly to zero. At point C, the equilibrium moisture content of the yarns is reached. At point B and C the adsorbed (absorbed) moisture evaporates.

Further drying of the filament is not advisable for the following reasons:
- yarns brushed at equilibrium moisture content have better physical and mechanical properties (properties deteriorate with overdrying);
- overdried filament tends to get natural (equilibrium) moisture from the environment, so energy consumption for overdrying filaments is not economically favourable.

The yarns must be fed from the sander to the dryer with a low moisture content so that the difference between the initial cw moisture content and the final cw moisture content is small.

Below is shown the equilibrium moisture % of fibre materials at shop temperature and humidity conditions φ =65% and {=20OC.

Acetate		thread-6	.0
Copper ammonia	silk-12		.5
Natural		silk-9 .5
Cotton fibre-		8.0
Flax fibre	8	.5
Polyacrylonitrile fibre		1.0
Polyamide	fibre-4		,0
Polyester	fibre-0		.5
	Polyurethane-1 .5
Viscose fibre		13.5
Wool fibre-14		.5

Drum sander machines, the most productive and rational in warp filling, with 16 drying drums evaporate moisture from warp yarns up to 1000 kg/h.

The design of the drum surface with anti-adhesion material and the stepwise

59

heating of the drum with temperatures from 80° to 120° improve the drying conditions and extend the range of the machine.

Specific steam consumption is 1.4-1.6 kg per 1 kg of evaporated moisture.

Due to heat reuse and thermal insulation of the drying section enclosures, the specific steam consumption can be reduced to 1.1 kg.

Moisture content in the removed air depends on the temperature of drying drums, at $t=120^{\circ}$ drum is 70 gr/kg, and at $t=140O$ - is 100 gr/kg, and as we know from the drying temperature depend on the cost of sanding.

The temperature limit for cooling the substrate on the first drying drums is more than 80°. With forced air blowing with dry preheated air up to 80° this limit can be reduced to 60° as the temperature reduction in this zone depends on the intensity of the air exchange.

The drum surface air blowing system has the following advantages:
- optimum utilisation of the waste heat in the heat exchanger;
- 20-30 % reduction of the temperature limit of the sanded substrate;
- using a low-power fan for blowing;
-constancy of climatic conditions in the workshop, due to the use of heat-insulated enclosures for the drying part of the sanding machine.

Machines with a horizontal drum arrangement allow the use of sanding machines with a large number of sanding baths and drums.

In convection dryers, only hot water heated by steam or electricity is used for yarn drying and a powerful fan is used for circulation. The evaporation capacity of the machine is up to 300 kg/hour.

Radiation dryers have heaters with radiation temperatures of 200° and higher. They have the lowest efficiency due to the high energy costs of production.

3.5. Pricing and winding devices.

The sanded and dried warp is fed to the price device, where the glued yarns are separated from each other (dry separation) by means of separation bars. Dry separation can cause damage to the surface of yarns, due to the forces F arising during the separation of yarns in front of the bars, due to the glued yarns, and the distance l to the bars B, the smaller, the greater this force and vice versa (Fig. 36).

Figure 36. Pricing device.

Dry filament separation is characterised by:
- Over-dried yarns require more effort to separate than yarns with normal moisture content;

- The vices of sanding can be determined by observing where the yarns separate;
- the number of separating bars is equal to the number of mending rolls minus one, as the first bar divides the threads into even and odd rows of threads, and the other bars divide the threads coming from each roller;
- large warp separation angles at the bar give high forces during yarn separation.

There is a close relationship between warp yarn separation forces and the physical and mechanical properties of the yarn, with decreasing yarn separation forces improving yarn properties and vice versa.

Yarn damage during separation can be eliminated by sanding with pre-drying and wet separation before final warp drying. Table 9 shows the results of the measurement of warp tension for yarns of linear density 11.8 tex and other parameters with and without pre-drying of the warp in the process of warp separation (Fig. 36).

Table 9.

Results of the warp tension measurement.

№	Parameters	Unit of measurement	Without pre-drying the warp in the yarn separation process	With pre-drying of the warp during the yarn separation process
1	Tension of filament branches up to the bar, F	sH	44,6	40,6
2	Filament separation angle, a	grad.	52	11
3	Angle of girth of the threads of the bars, β	grad.	39	20
4	Total separation force, Fo	H	320	50

The analysis of Table 9 shows: pre-drying of the warp reduces the yarn separation angle a, and this causes a decrease in the yarn girth angle ß and in the tension of the branch coming off the bar F, resulting in a decrease in the total force F_o of the warp yarn separation.

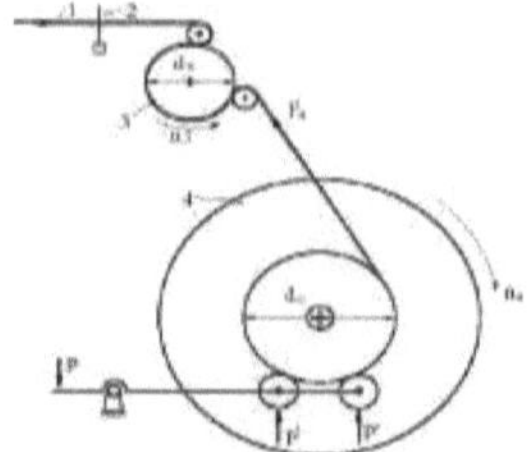

Fig.37. Winding device.

After passing the price device, the warp yarns 1 pass through the width-adjustable sliding row 2, round the outlet roll 3 and are wound on the weaving warp 4 (Fig. 37), by means of a winding device.

In order to obtain the same tension in the area of the outlet roll and the weaving head, it is necessary that the speed of the thread coming off the roll 3 (v_3) should be equal to the speed of thread winding on the head (v_4), i.e. $V3=V_4$, and since V πῖπ, we obtain:

π- d $_3$- n_3 = $_{\pi\text{-}á4\text{-}n4}$; since $n_3 / n4 = d_4 / d_3$

When increasing the winding diameter d4 on the warp, the speed of the weaving warp must be reduced, otherwise the warp tension in this area decreases and eventually the warp breaks.

There are several systems to drive the winder or to control and regulate the tension of the threads during winding - transmission, hydraulic and multi-motor.

The transmission system is widely used due to its reliability and ease of maintenance, low cost and high drive power. The system has a drive with one main motor providing all motions of the sander organs and a clutch with any load (pneumatic, electromagnetic, etc.) limiting the action of the planetary differential mechanism to regulate the input torque. The automatic speed and torque control of the weaving slider is performed during the entire winding run on the slider.

The hydraulic systems use pressurised fluid to provide constant power (high speed at low torque and low speed at high torque) to the weaving slurry. The complex design makes fault detection difficult, the simple connection (not mechanical) of the components to the slurry drive shaft ensures reliable operation of the drive.

In a multi-motor system, all the working parts are driven by DC action and the warp is driven by a separate DC motor, which ensures that the warp tension remains constant as the warp diameter increases on the warp. The system has a wide range of power and warp winding mode on the warp head. The winding device can be used to wind a single warp, two side-by-side or overlapping warps, a reverse warp rotation system that allows the warp to be wound on left and right warps, or a warp winding sealing system

3.6. Control and regulation of sanding parameters.

Control and regulation of the sanding parameters ensures the production of high quality sanded bases.

The main parameters include sanding speed, squeegee roll pressure, warp tension by zone, temperature conditions of sanding and drying, warp moisture, glue, sanding concentration and drafting.

By means of a microprocessor, these parameters can be recorded and precisely controlled.

The use of automatic control systems for the sanding process frees operators from a number of standard parameter measurements and allows them to adjust: loading of the braking systems of the rewind rollers, to ensure the specified tension of the threads; the level of the slurry in the glue trough; the temperature of the slurry in the trough; the pressure of the squeezing rollers; the speed of the spinning process; the immersion of the threads in the slurry; the temperature and speed of the drying drums; the tension of the threads on the warp; the moisture content of the warp; the density of the winding on the warp; the change in the length of the winding on the warp.

In addition, the control system performs the following operations - it calculates the yarn weight on the spinning roller, determines the expected yarn unwinding end time, keeps a record of the full warp run, accumulates data on the conditions of spinning each metre of warp (speed, tension, temperature, etc.), deviations from the set values are printed out and attached to each warp.

The computer then receives data on the efficiency and processing quality of each metre of warp on the weaving machine. By reporting defective spots on each warp, the system ensures the efficiency of the weaving machines.

The sanding machine is equipped with measuring devices and regulators to ensure high quality of sanded substrates.

The principle of operation of the electronic warp drawing device is based on the following: per unit of warp length (1 metre), a certain number of pulses are produced. The elongation or shrinkage of the yarns a is expressed as the difference between the number of impulses on the warp gn and in the sanding bath gv, divided by the number of impulses in the sanding bath gv and multiplied by 100%, i.e.

$$a = \frac{r_n - r_a}{r_a} \cdot 100\%$$

The measuring cycle depends on the speed of the machine, i.e. the faster the machine runs, the more often the values of the measured parameter are recorded. The speed of the machine is digitally recorded using a sensor on the nava.

By connecting the drafting meter to the job computer, it is possible to achieve a specified thread draft in the different zones of the sander. Table 10 shows the recommended tension for the different zones of the sander.

Table 10.

Recommended tension by zone of the sanding machine.

№	Sanding machine zones	Recommended tension, % of yarn breaking load	Single thread tension 25 tex, sN
1	Rack - first pulling shaft	1-2	7
2	First draw roll - squeezing rolls	1-2	7
3	Drum - second pulling shaft	2-3	10
4	Price field	6-8	28
5	Exhaust shaft - navy	10-12	42

In the electromechanical warp drawing device, the actual tension is sensed by a spring-loaded pendulum connected to a potentiometer, which regulates the state of the equalising mechanism by means of a servomotor until the actual tension stabilises with the set warp tension.

By adjusting the machine speed depending on the residual moisture content of the ground yarns, the temperature on the surface of the drying drums is constant, so the time of the yarns passing through the drying drums and therefore the speed of the yarns determines the degree of drying. A sensor at the outlet of the drying drums records the actual value of the warp moisture, which is compared with the set value of the warp moisture in the electronic moisture device. If the actual warp moisture value deviates from the set value, impulses are generated from the electronic warp moisture device to change the sanding speed. The speed is stabilised when the actual and set warp moisture values are the same.

The precise temperature control of the drying drums is ensured by an electronic device that supplies current to a converter that transforms this electronic signal into a pneumatic one, the latter controlling the valve of the vapour pressure regulator.

To control and regulate the parameters in the glue bath, electronic devices are used to control the pressure of the squeezing rollers, the temperature of the slurry and the level of the slurry in the bath.

Electronic controllers change the pressure force in the squeezing rollers depending on the speed of sanding, which allows to obtain the same degree of threads sanding both in quiet running and in working running of the machine.

The electronic slurry level device automatically maintains a continuous supply of slurry into the glue bath and prevents the slurry from thickening in the slurry line. Continuous feeding of the slurry is realised by means of a slurry line in which the volume of the slurry passing through is less than the volume of the consumed slurry. Slurry level sensors can react to differences in the slurry level or to temperature differences between the ambient and the slurry medium.

The moisture content of the warp yarns can be measured directly by: capacitance of the capacitor between the plates of which the warp is moving; electrical resistance of the yarn. Wet yarns have a lower electrical resistance and dry yarns a higher one. The received impulses are fed to the electronic device, which signals the speed control motors to increase the sanding speed when the warp moisture decreases, or to decrease the sanding speed when the warp moisture increases.

The above described electronic devices perform narrow control, i.e. a separate controller for each control loop (level, temperature, humidity, etc.).

During the grinding process, it is possible to use switches, keypads or data carriers (magnetic cards or tapes) in which centralised parameter setting is entered.

When combining all control and monitoring functions in one device we will get a computing device (microprocessor), which is designed specifically for controlling a given technological process. In addition, the development of analogue-to-digital converters (ADC) with appropriate switches, optically connected input and output circuits and sensors (temperature, humidity, etc.), controlling the conditions of the process and binding (ADC) microprocessor to a specific technological process largely solves the problem of automation of the polishing process.

In the polishing process, the yarn properties change: yarn weight increases due to the gluing and thus the yarn linear density increases; yarn strength increases significantly due to the gluing of the individual fibres; yarn elongation decreases as the glued fibres do not slip relative to each other; yarn durability increases due to the protective film of the polish.

Yarn strength increases by 20-25% after sanding, abrasion resistance increases 5-10 times, elongation decreases by 20-30%.

3.7.Specialised sanding equipment.

Special sanding machines prepare dyed backings to produce low-maintenance denim and jeans-type garment fabrics.

There are - indigo dyed sanding, dyed sanding and resin sanding.

Slitting with indigo dyeing. The traditional method of preparing indigo dyed warp is as follows: knotting of yarns (grouping) into a ball; dyeing of bundles obtained from the balls; washing and drying of the bundles; batch knotting of the bundles; sanding on drum machines.

As can be seen the process of dyeing basics is very labour and capital intensive, requiring skilled personnel.

Indigo base dyeing machines are manufactured by Zukker and West Point. They include: a rack for the re-sewing rolls, one or two bars for soaking or pre-dyeing and washing; four or five dyeing bars, including air cooling above them; one or two washing bars for washing out loose dyes and chemicals; a drying drum for pre-drying; a compensator to take up the warp when the machine stops; a sanding machine with one or two baths.

The choice of the construction of the sanding machine is determined by the technological parameters of the filling. Two sanding baths and thirteen drying drums are used for preparing the warp for dense denim. The process requires a very high quality of warp preparation. There must not be any broken threads on the re-sanding rollers, which can lead to the formation of clamps on six-metre-high teddies.

The cost of sanding machines is high, so high quality indigo dyed preparation of the spinning rolls represents a significant step forward compared to the old indigo dyeing system.

Flaking with dyeing. The increase in the production of denim and denim fabrics with indigo shortages has stimulated the development of dyeing in the sanding process. In addition, dyeing with sanding is much faster than indigo dyeing.

Initially, a dyeing machine with a single glue bath was used, into which the dye solution was first fed. The resulting woven warp was then rewound and the chemicals required for etching and sanding were poured into the glue bath. Large machine downtimes and contamination of the glue bath resulted in a reduction of machine productivity.

The installation of a second glue bath and a pre-drying unit between the baths allowed the mordant and the sand to be fed at the same time, and the substrate to be fully painted and sanded.

For very dense fabrics, such as denim, it is possible to dye separately in parts in two baths, followed by pre-drying of the dyed fabrics, and in the third bath the fabric was treated with mordant and polish and then the whole dyed and polished fabric was dried.

The production of striped denim with two colours and dyed yarns was carried out using four baths. The first and second baths are fed with dyeing solution, the third with mordant and schlichta, and then the two-colour dyed and polished warp goes to the drying drums, a part of the dry warp goes to the fourth bath is polished, pre-dried and goes to the general drying part of the machine.

Stitching machines with 24 drying drums and gluing baths are equipped with compensators to prevent stripes on the substrate during short stops of the machine. The compensator consists of a series of mounted movable and fixed rollers that automatically select the substrate after drying when the machine is running quietly.

Schlichting with dyeing ensures high uniformity and colour strength compared to indigo dyeing!

The composition of the dye solution includes dye (naphthol, sulphur), caustic soda, methyl or ethyl alcohol, alginite, wetting agent, defoamer, and in the etching and sanding solution - starch, caustic soda, acetic acid.

Resin slitting. Offers the simultaneous application of a thermosetting resin with a catalyst, a dye and a bobbin on the yarn in the same bath. The resin materials, due to polymerisation, become an integral part of the fabric. After weaving, absorption (absorption) of the dye takes place during heat treatment of the fabric. The fabric is then saponified, washed, dried and additionally treated with resin (for weft yarns).

Advantages: low degradation of cellulose fibres; reduced capital investment; use of cheap pigment dyes; high durability of colouring; low sublimation (transition from solid to gaseous state when heated); minimal water consumption; simple process.

Distillation machines are used for the distillation of unsplit warps from the spinning rolls to the weaving warp. In some cases, the twisted yarn is treated with paraffin during the distillation process. The distillation speed is up to 70 m/min.

Distillation-emulsifying machines are designed for emulsifying and winding cotton and silk twisted yarns, woollen yarns and their blends with chemical fibres on the weaving warp from the spinning rolls or weaving warp. The moisture content of the warp after emulsification is 15-30% and the warp distillation speed is up to 80 m/min.

4. PIERCING AND TYING THE BASES

The final operation in preparing the warp for weaving is picking or tying. Picking as a technological stage includes the insertion of the warp threads into the lamellar eyes, gules and reed teeth. It is carried out in case of a change in the range of fabrics produced, which entails a change in the warp threading of the weaving machine, including: the number of warp threads, the reed number, the number of reams and the sequence of threads in the galleys. Wear and tear on reeds, lamellae, reams (dobby frames) also causes the need for threading. In weaving production, on average no more than 10-15% of the total number of warps are picked. Less labour-intensive and more widespread than picking is warp tying, which consists in knotting the ends of the yarns of the reworked warp with the ends of the yarns of the newly prepared warp. The tying can be done directly on the weaving machine or in the picking department. After tying the yarns together, the new warp is pulled through the lamellas, galeva and reed. The joining of the yarns of the new and finished warp by twisting and gluing the ends of the yarns is called piecing and belongs to the knotless joining methods. Piecing is used when the passability of the glued areas is higher than that of knots. For example, in cloth weaving at high yarn line densities (360 tex and higher), in complex remise and jacquard dressings.

The lamella is a part of the warp monitor mechanism designed to stop the weaving machine automatically when the main thread breaks. Depending on the principle of operation of the warp monitor, the lamellae are made of the following types: L - closed form, used in mechanical action mechanisms; LO - open form, having a through slot on one side and used in mechanical action mechanisms; LE - closed form, used in electric action mechanisms; LOE - open form, having a through slot on one side and used in electric action mechanisms. The size of the lamellae and their weight depend on the linear density of the warp yarns. As the linear density of the yarn increases, lamellas of higher weight are used.

Remizki (remise frames) consist of the frame and the woofers. The main dimensions of the weaving frames are: the frame swing, which should always be 1.5-2 mm less than the swing of the wefts, which is necessary for some mixing of the wefts together with the threads in the horizontal direction, the height and width of the frame. On weaving machines, metal (from remise wire) and lamellar galleys are used. Depending on the brand of the weaving machine and the produced assortment, galleaves are used with a height from 265 to 710 mm and an eye size from 3.2 to 12.0 mm in length and from 1.5 to 6.0 mm in width. Increasingly widespread use are plate galewa, which are made of two types: I -

for the production of silk and cotton fabrics, II - for the production of technical fabrics. Well-polished lamellar galleys allow to reduce the breakage of main threads during shedding in comparison with galleys made of remise wire.

The reed is designed for even distribution of warp yarns across the width of the fabric and to regulate the density of the fabric on the warp, to bring the weft yarns to the fabric underlay and is one of the guides for the shuttle or weaver. The reed number is the number of teeth per one decimetre of working width of the reed. The reed number can be calculated by the formula

$$N_6 = \frac{P_0\left(1 - \frac{a_y}{100}\right)}{Z_\Phi} \qquad \text{(зуб/дм)}$$

(tooth/dm)

where: P_o - WARP YARN density; Au - weft yarn yield; Z_Φ - number of yarns penetrated into the reed tooth.

The correct choice of the reed number determines the warp breakage during fabric production and the evenness of the warp yarns in the fabric. The warp breakage depends to a large extent on the filling of the gap between the tines. The knots on the warp threads must be able to pass freely through this gap.

4.1 Getting the basics right

Picking can be done manually, on semi-mechanical machines and on automatic machines. Two workers, a picker and a feeder, pick up the warp manually on a nimble machine. The feeder picks out the warp threads coming from the warp and feeds them to the hook, which is threaded through the galewa eye by the picker. If closed lamellas are used, the threads are first passed through the eyes of the lamellas and then into the galeva. After threading a set number of warp threads through the eyes of the lamellas and galleys, the sorter pulls them between the reed tines with a passet.

The passet is a thin, slightly curved, oval-shaped plate with two slits at the tops of the oval. During operation, the passet is rotated 180° , and the threads inserted in its oval are pulled into the reed tines. At the same time, the other diametrically located oval passes into the adjacent reed tooth and the passet automatically moves one tooth along the roll. The picker and the feeder pick a maximum of 650 to 1200 threads per hour (depending on the type of picking, reed number and qualification of the workers). Semi-mechanical picking machine PS is operated by one picker. The machine is designed for mechanical selection of warp yarns, mechanical picking them in the reed and manual picking in the eyes of the galev. Selection of warp yarns is carried out both with and without laid prices. The speed of filament feeding to picking is up to 100 per minute, the productivity of the machine is regulated by the speed of manual picking in the galeva and depends on the labour skills of the picker.

The Barber-Kolman automatic picking machine (USA) is designed for automatic picking of single-coloured or coloured bases from one or two sheds into the lamellae of galev and reed. The automatic machine consists of a frame with a movable carriage and two mobile trolleys for the installation of bases - working and spare. The sequence of operation of all mechanisms is controlled by punch cards. The machine is driven by an individual electric motor. The speed of picking on the working stroke of the machine 140 threads per minute, on a quiet - 20 threads per minute. The productivity of the machine is 4000-5000 threads per hour, with complex picking - 3500 threads per hour. The machine can pick through threads in 26 lamellas and up to 6 lamellas in one filling. The technological scheme of the picking machine is shown in Fig. 38б. From the electric drive *1* (Fig. 38a) through pulleys *2* and 3, gears *4, 5 and 6*, bevel gears *7, 8* rotate cams *9*, which through rollers *10* transmit motion to the levers *11*. Through the toothed sector *12* and gear *13* rocking motion is transmitted to the single arm lever *14*. The latter is connected to a flexible needle *15* placed in a guide *16*. The picking operation starts when the needle *15* passes through the reed tooth, the eye of the selected lamellar galeva and the lamella.

The gap between the reed teeth, necessary for the needle to pass through, is created by the action of the slider *17*, mounted on the roller *18*. The disc of the slider with a diameter of 50 mm with its profiled plane increases the gap between the teeth. After picking through the thread, the slider, placed on the carriage, moves perpendicular to the movement of the needle. During the picking process, the gale is picked out. On this machine, the threads are picked through only the slatted galleys (Fig. 38c). At the upper and lower ends of the galeva have open holes *1* for putting on the remise bars. At a set of galevos on strips at first a galevo G₁ is put on, the hole *3* of which lug is directed downwards, and then a galevo G_2 with the lug directed upwards. It is necessary for consecutive selection of galley by the galley picker. The warp yarns are taken through hole *2*. The lamellae L_1 and L_2 (Fig. 38g) have similar holes *4* for the lamellae pickers, the round holes 5 being for the warp threads. Before the needle enters the eye of the gallevo and lamella, the gallevo picker *19* (see Fig. 38b) and the lamella picker *23* select the next gallevo *20* and lamella *22* and feed them into the area of the gallevo feeder *21*, along the grooves of which the gallevo and lamella move. When coming off the screw groove, the galevo and lamella are also rotated by 90° and set perpendicular to the moving needle *15*.

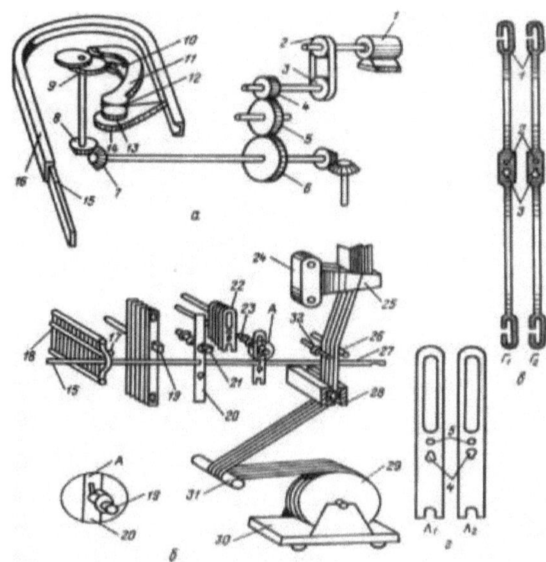

Fig.38. Technological scheme of the picking machine.

Selection of the main thread from the total warp web is carried out by one or two worm-shaped yarn pickers *26* and *32*, which can be operated from one or two navoi. The navoi *29* are placed on a trolley *30*. The warp yarns from the navoi circle the guide roller *31, and are* clamped in clamps *28* and *25*. The latter is mounted on a moving bracket *24*, which is used to adjust the warp tension in the filling.

When picking up the thread, the needle *15* with its hook *27* grabs the thread and returns back, pulling the thread through the eyes of the lamellae, gale and reed tooth. The operation of the device machine is controlled by punch cards, which are made on a special cardboard secal machine. The company "Zelweger-Uster" (Switzerland) produces probornyy machine EMV. It is designed for proborki warp threads in the gallevo, reed and lamellas of any type from one or two navoi at a speed of 150-160 threads per minute with the number of remizoka from 2 to 28. In addition to instrument machines, the company "Zellweger-Uster" has developed machines for thread picking in galeva remizok and set of lamellas (without picking in the reed) from one or more navoi. An additional machine is used for reed picking. This two-stage system is more versatile, as it does not require special equipment for the weaving machine guides. It can be used to pick up yarns of any height, both twisted and lamellar. The maximum picking width is 400 cm. As a result, automatic picking can be used on wide weaving machines.

4.2 Tying the bases

The automatic tying of the new warp yarns to the yarns of the finalised warp is carried out by knotting machines. A distinction is made between stationary, mobile and universal machines. Depending on the method of thread selection, knotting machines are divided into machines with needle, price and combined selection. All knotting machines are labelled with numbers (125, 190, 200, 250), which indicate the maximum filling width in centimetres.

UP1-5 machines with different working widths have needle yarn selection, UP2-5 machines have price selection, and UP-6 machines have combined selection. The knotting speed reaches 500-600 knots per minute and is set depending on the linear yarn density, type of fibre, yarn density in the warp.

Machines with needle selection are used in the cotton industry, with price selection - in silk, wool, and also in the cotton industry when tying multicoloured warps. Stationary knotting machines tie the warp in the warp section. When the warp is being finished, the weaving machines are used to remove the lamellae, heddles and reed together with the ends of the old warp, which are tied in knots, and a 10 cm wide strip of fabric is left on the side of the reed (so that the threads do not come out of the removed working parts). All of this is transferred to the probornoy department and set on the knotting machine, where the ends of the threads of the old warp are tied to the ends of the threads of the new warp, and the knots are dragged through the lamellas, galeva remizok and reed. The stationary knotting machine consists of five basic elements: two mobile trolleys for transporting and installing the new warp. After the preparation of the warp, the trolley on rails is brought to the knotting machine, and on the second trolley at this time prepare the next warp, which at the end of tying immediately installed on the knotting machine, thus reducing machine downtime;   preparatory machine-charger, designed for the
preparation for tying the warp removed from the machine, where the threads are parallelised and clamped; the upper mobile carriage used to clamp the prepared old warp and move it to the root machine; the root knotting machine, on the guides of which the knotting mechanism moves, where the tying of threads is carried out; the knotting mechanism, taking and tying the ends of the threads of the old and new warp.

Mobile knotting machines tie the yarn ends of the finished warp to the yarn ends of the newly threaded warp directly on the weaving machine. Fig. 39 shows the technological scheme of the mobile knotting machine UP-2M.

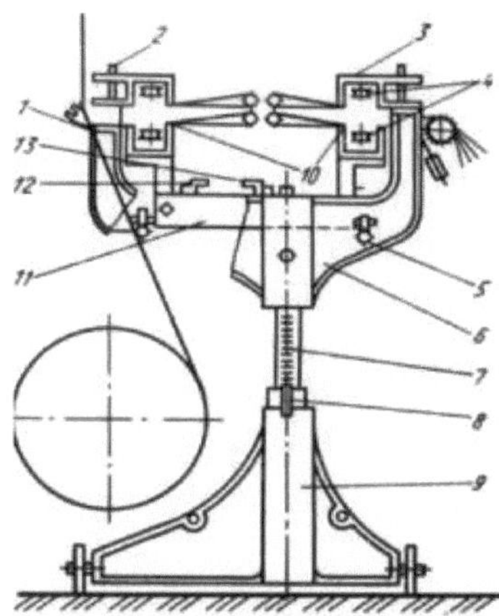

Fig.39 Technological scheme of mobile knotting machine

The mobile stand with clamps consists of two pairs of brackets, lower *9* and upper *6,* connected with each other by toothed posts 7, which allows setting the clamps to the required height with the help of the handle *8* when threading at weaving machines. Removable boxes of upper clamps *3* are installed on the upper brackets, connected with the brackets by means of pins 2 passing through the hole of the clamps' legs. The lower clamps *10* are movable relative to the upper ones, as they are attached to two brackets *11 with* wheels resting on guides 5 fixed on the upper links of the rack, and can move along them. The brackets *11* are linked together by ties and constitute a rigid frame to which a toothed rack *12* is attached. A fixed rail *13 is* attached to the rack brackets 6. The worms of the knotting carriage engage with both rails.

The carriage receives a progressive movement along the stationary rail *13* from right to left as the tying process proceeds. By means of the rail *12,* the lower clamps with the new warp / can be moved along the guides 5 relative to the upper clamps to one side or the other, so that the outermost next thread of the new warp is positioned exactly under the outermost thread of the old warp. The upper and lower clamps are arranged in the same way and have the form of boxes or troughs of rectangular cross-section with rubber gaskets on the side walls and wooden blocks on the bottom. Metal linings *4* are placed on these blocks. Each pad consists of two strips, the contacting surfaces of which are made in the form of teeth. As a result of displacement, the two strips press with their ribs forcefully against the side walls of the clamp boxes and tightly press the base, which runs across the clamps under the plates *4.*

First, the new backing is tucked into the lower clamps *10, having* first removed the upper clamps *3,* then the old backing is tucked into the upper clamps, having placed them on the lower clamps. The base is carefully straightened, parallelised manually with brushes and fixed in the clamps in a taut position. On the upper clamps set the knotting carriage, pass the price cords of both bases through the price tubes of the carriage, hitch it with worms with laths *12* and *13*. Using the manual drive, check its operation by tying several knots, and then switch on the electric motor.

As the threads are tied, the carriage must move forward and the lower fixed clamps must move so that the next outermost thread of the lower warp is positioned exactly under the next thread of the upper warp. For this purpose, the ends of the feeler gauges *1* and *6* (Fig. 40a) are positioned above and below the end threads *7 and 8 of the* lower and upper warp. Both make a rocking movement from the cam *9*. The springs ensure constant pressure of the rollers against the cam *9*. Oscillation of the feelers is symmetrical with respect to both bases, i.e. they alternately converge and diverge and through the articulated with them levers *2 and 3* cause switching on and off the feed dogs *4* and 5, acting on the teeth of ratchets *10* and *16* (Fig. 40b). The dogs make reciprocating movement in horizontal direction from two cams *18* and *19* through levers *17* to *20*. Dogs *13* are retaining dogs.

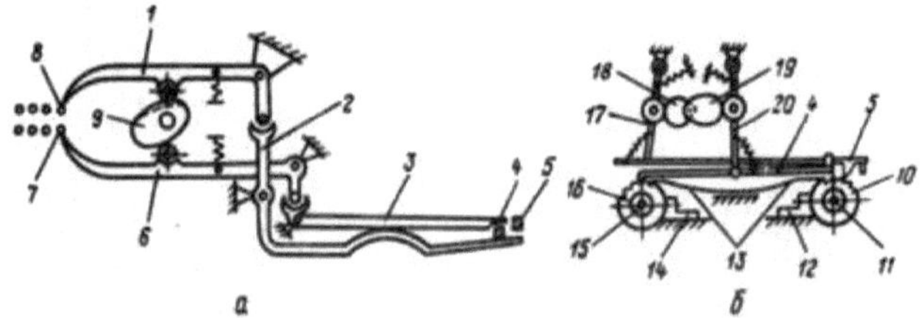

Fig.40. Schematic diagram of the knotting carriage.

The ratchet *10* is made as one unit with a worm *11* engaging with the rail *12 of* the upper stationary clamps, and the ratchet *16 is* made with a worm *15* engaging with the rail *14 of* the lower movable clamps. Thus, the upper stylus is connected with the fixed upper clamps and the lower stylus with the movable clamps. Movement of the carriage takes place only when there is no
another thread *8* (see Fig. 40a*)* under the upper stylus or when there is no another thread *7* under the lower stylus. There are four possible cases:

1) if the next threads of both bases are present and correctly positioned, the ends of the styli will bump against the tensioned threads and will not make a complete counter movement, as a result of which both feed dogs *4* and 5 will be switched off and the carriage and lower clamps will not move;

2) in the absence of the next thread of the upper warp, the stylus will make a full trip and lower, as a result of which the dog 5 will turn the ratchet *10* and the carriage will move forward; since there is a thread *7* of the warp under the lower

stylus *6*, the worm *15* will not turn, but will stop in the teeth of the rail *14* and the lower movable clamps will move together with the carriage;

3) in the absence of the next lower warp thread, the lower stylus will travel all the way round and allow the dog *4* of ratchet *16 to* engage, so that the lower warp clamps will move towards the carriage, as the latter remains stationary in the presence of the next upper warp thread;

4) in the absence of threads of both bases, both feed dogs *4* and *5* are switched on and the carriage moves forward to the threads of the upper and lower bases, in this case the action of the worm *15* on the movable rail *14* is neutralised by the carriage moving forward.

Selection of the next threads for tying is performed by the joint action of two carriage mechanisms: price separators and picking brushes. The price separator mechanism selects the next threads one by one from the old and new warp by shifting the prices of the individual threads away from the carriage (Fig. 41). The two threads separated in this way are caught by two picking brushes and moved aside. The trajectory of the picking brushes is shown in Fig. 41a.

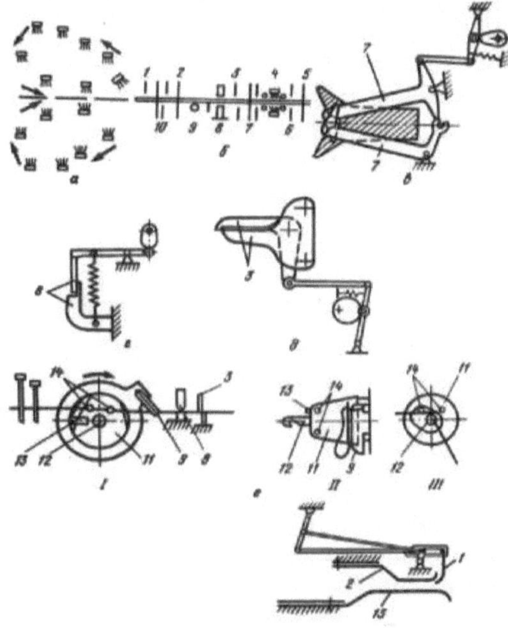

Figure 41: Mechanism of price separators.

The selected yarns are placed in the working area in the following sequence (Fig. 41 b): to the left of the brushes *4* there are front folding levers *7*, scissors *3*, clamps *8*, knotter beak *9*, dropping lever *2*, rear folding levers *10* and hook *1* for picking up bound yarns; to the right of the brushes there are feeler levers 6 and waste hook 5. The front folding levers *7* (Fig. 41 c) engage and bring the yarns closer together, which are clamped by clamps *8* (Fig. 41d) and trimmed by shears *3* (Fig. 41 e). The rotating hook 5 then takes the cut ends of the threads to the waste. The clamps 8 are opened and the threads, captured by the clockwise rotating beak 9 (Fig. 41e), are wound on the head of the tube *11*. At the same time the beak, moving backwards, passes with the captured threads behind their left ends. Then, with further rotation and forward movement, it places the threads on top of their left ends and between the horns *14 of* the tube, forming a loop. By this time, the tightening needle *12* comes out of the tube and with its hook grabs the threads lowered by the beak, pulls them into the tube and holds them there. After that, the wiper *13* will move forward, drop the loop from the head of the tube, by the joint action of the wiper lever *2* (Fig. 41j) and the take-off hook *1* tighten the knot and leave the area of its formation. Different needles are recommended for yarns with different line densities.

Universal knotting machines can be used as mobile or stationary knotting machines depending on the specific application. They are conventional mobile knotting machines equipped with two clamping stands that bind the warps in the instrument department. The presence of two racks allows a more rational use of the knotting head, i.e. on one rack the yarns are tied, on the other the next warp is prepared for tying, thus increasing the productivity of the knotting head. Universal knotting machine UP-6 is designed for automatic binding of cotton, woollen, linen yarns, silk and chemical yarns. The machine is equipped with a needle-price mechanism of yarn selection, which allows using different methods of selection when tying bases: needle, price and combined (one of the bases with a "price cross", the other without it). The quality of the knots on the UP-6 machine ensures that they can pass through the loom guides without loosening. The length of the ends of a double knot is 0.5 mm longer than on the UP-2M and UP-5 machines. The number of defects during tying is reduced by 1.3-2.1 times.

Modern knotting machines used in the domestic industry bind 300-400 knots per minute. The actual productivity of the machines depends on the downtime associated with the preparation of new and old warps for tying. Depending on the type of warp, linear density of yarns and the number of yarns in the warp, the actual productivity of stationary and universal machines varies from 8000 to 12000 knots per hour.

Mobile knotting machines operate at a lower capacity, as their use increases the time required to prepare the warp for tying. The actual capacity of these machines is up to 3000-8500 knots per hour.

Modern knotting machines are equipped with a whole range of monitoring devices. For example, on machines from Titan (Denmark), the tying of threads is controlled automatically: if a couple is detected, the machine stops.

The performance of the knotting machine is as follows:

1. $\Pi_{\phi y} = \dfrac{n_{\text{гл}} \cdot 60}{2} \cdot \text{КПВ}$ Capacity of units per hour

2. $\Pi_{\phi 0} = \dfrac{n_{\text{гл}} \cdot 60}{2 \cdot n_0} \cdot \text{КПВ}$ Capacity of bases per hour

3. $\Pi_{\phi} = \dfrac{n_{\text{гл}} \cdot 60}{2 \cdot n_0} \cdot G \cdot \text{КПВ}$ Capacity in kg per hour

where: $n_{\text{гл}}$ is the number of main shaft revolutions min; n_0 is the number of warp threads on the warp;

G - weight of the weaving bulk; KPV - useful time coefficient.

Carbon monoxide is determined by the following formula

$$Y = \frac{\ell}{L_{\text{H}}} \cdot 100\%$$

where: i *is the* length of the ends going into the yarns; L_{H} is the length of the warp yarns on the warp.

Vices: missing reed tines; broken pattern; pairs; missing threads, two threads caught, etc.

LITERATURE

1. Prabir Kumar Banerjee. Principles of FABRIC FORMATION. © 2015 by Taylor
& Francis Group, LLCCRC Press is an imprint of Taylor & Francis Group, an Informa business.
2.Nikolaev S. D et al. Theory of processes, technology and equipment of weaving production. M., Legpromizdat, 1995. 256c.
3.Ormjord A. Modern preparation and weaving equipment. M, Legpromizdat., 1987. 211c.
4   . G. H. Oelsner's. A Handbook of Weaves is the best known and most accessible weaving pattern book. November 18, 2004.
5   Talavashek O, Svatyi V. Threadless weaving machines. M., L.I., 1985, 334 pp.
6   . K. L. Gandhi. Woven textiles Principles, developments and applications © Woodhead Publishing Limited, 2012.
7   Fronczak I., Wnuk J. Weaving, ch. P, Warsaw, 1978, 326 pp.
8   Rakhimkhodjaev S.S., Kadyrova D.N. Theoretical bases of the process of tissue formation. Textbook. Tashkent. TITLP. 2018.
9   Simon L., Hübner M. Technology of yarn preparation for weaving and knitting production. M., Legpromizdat, 1989 270 pp.
10 . HANDBOOK OF YARN PRODUCTION Technology, science and economics P R Lord, NCSU, USA 504 pages 244 x 172mm hardback July 2003.

yes
I want morebooks!

Buy your books fast and straightforward online - at one of world's fastest growing online book stores! Environmentally sound due to Print-on-Demand technologies.

Buy your books online at
www.morebooks.shop

Kaufen Sie Ihre Bücher schnell und unkompliziert online – auf einer der am schnellsten wachsenden Buchhandelsplattformen weltweit! Dank Print-On-Demand umwelt- und ressourcenschonend produzi ert.

Bücher schneller online kaufen
www.morebooks.shop

Printed by Books on Demand GmbH, Norderstedt / Germany